Pelican

T0283907

Animal

Series editor: Jonathan Burt

Pelican

Barbara Allen

REAKTION BOOKS

For the pelicans, and for all researchers, ornithologists, conservationists, writers and volunteers who work to protect these magnificent creatures

Published by
REAKTION BOOKS LTD
Unit 32, Waterside
44–48 Wharf Road
London N1 7UX, UK
www.reaktionbooks.co.uk

First published 2019
Copyright © Barbara Allen 2019

Printed and bound in China

A catalogue record for this book is available from the British Library

ISBN 978 1 78914 075 0

Contents

Introduction

There are certain creatures that get under, or align with, our emotional skin. Although I love pigeons, and have several white doves as members of my household, years ago pelicans made their way into my heart and soul. From a young age, I would squeal with delight at their large bodies and short legs, able to achieve so much, which appealed to my non-athletic, asthmatic being. I took on its physicality, feeling 'pelican like', with large belly while pregnant; I too enjoyed swimming, my weight cupped gently by the water's buoyancy. A wooden misericord of a pelican adorns the mantelpiece, a solemn reminder of long hours in ministry, when pastoral demands mirrored the legendary sacrificial nature of the pelican.

Wild, mostly silent, dignified. At times caricatured, or lampooned, these magnificent birds have rarely been tamed; they are not a bird for the drawing room, except within the pages of Edward Lear's books, via his lavish illustrations or his sparsely versed *Nonsense*. Who can bypass a pelican? Surely they illicit a response; they are too big and beautiful to be ignored.

And yet, as you will find within these pages, they have been ignored, to their peril and ours. They do not have the 'wow' factor of charismatic megafauna; they do not reside with us in homes, on hearths (except for the occasional Victorian taxidermy specimen, on eternal watch, in a museum), perches or beds. Yes, they

American white and brown pelicans, Florida.

are present in certain seaside towns, spotted cruising above coastal climes, or on lakes, but too often they are an afterthought, a bonus for tourists. And yet, if you wander into a cathedral, you may notice a pelican carving on a choir stall or embroidered on an altar cloth; stroll through St James's Park and acknowledge their presence, gifts to royalty. Pelicans bridge the sacred and the secular.

Groups of pelicans are known by several unusual collective nouns: a brief, a pod, a pouch, a squadron and even a 'scoop'. The word 'pelican' has been applied to a number of items that are not birds. 'Pelican' was an early dental extraction instrument, originating in the fourteenth century. Its lever and claw, which resembled a bird's beak, would be manipulated to prise a tooth from its socket. It is the name of a glass vessel, utilized in the science of alchemy. In battle, 'pelican' can refer to shot, or shell; it can also be a hook, in the shape of a pelican's bill. 'Pelican' can refer to a creature from a very different species; assassin spiders, also known as pelican spiders, have elongated jaws and necks designed for catching other spiders. But the most bizarre, perhaps, is the

Dental pelican for tooth pulling, Europe, 1701–1800.

term 'pelicanist'. This is not someone who is passionate about pelicans. The term 'pelicanist', coined by Jerome Clark, an American writer who specializes in the paranormal, is used to describe a person who dismisses reports of a possible UFO sighting by proposing other reasons for the object, including suggesting that the UFO was a flock of pelicans.[1]

Within these pages you will read about avian pelicans, not equipment, spiders or UFOs. My hope is that pelicans continue to grace the skies, coastlines and wetlands, and that their numbers increase. May they never become so rare as to be misidentified as UFOs.

Misericord at the Church of St Peter and St Paul, Lavenham, Suffolk.

9

THE PELICANS,

In the Gardens of the Zoological Society.

1 Biological Marvels

A pelican walks into a bar.
The bartender asks him, 'Why the long face?'

The biological story of the pelican is embedded within fossils, theories about evolutionary stasis, Guinness World Records and new discoveries about the nature of flight. The pelican is old, descended from one of the most ancient families of birds, but *pelican* should be thought of as plural, for there is more than one species. Until recently there were thought to be seven, but in 2007 the Peruvian pelican, which had been classed as a sub-species of the brown pelican, was recognized as a separate and distinct species, increasing the number to eight. Pelicans' biologies (and biographies) are open, not yet fixed; more is being revealed, including the devastating ways that human history has meshed with, meddled in and changed (and continues to change) breeding habits, habitats, food supplies and, at times, their very lives.[1]

We begin as far back as we can, with what has been unearthed to date. Fossils found in Eocene deposits, dated to 40 million years ago, are similar to the modern species of pelican.[2] There are problems, however, with the exact dating of the pelican, for the order to which it belongs, the Pelecaniformes, is large, traditionally thought to be comprised of six families and 65 species. Pelicans are closely related to the shoebill and hamerkop, with ibises, spoonbills and herons as distant relatives.[3] 'Evidence from DNA suggests pelicans may be more closely related to New World vultures and storks than to sulids and cormorants.'[4]

Thomas Kelly's coloured etching of two pelicans at the Zoological Society of London.

11

J.Smit

$\frac{1}{6}$

The oldest 'pelican' discovery is thought to be a femur from the late Eocene (40 million years ago), found in the Paris Basin and named *Protopelicanus cuvieri* by Ludwig Reichenbach in 1853. However, there is disagreement among researchers concerning whether it was a typical pelican, a Pelecaniform but not a pelican or a femur from a sulid.[5] Up until the last few years, the oldest undisputed pelican was *Pelecanus gracilis*, named by Henri Milne-Edwards, a French zoologist, in 1863, which came from the early Miocene (20 million years ago) in France.[6]

What is now believed to be the earliest known pelican fossil, dated to 30 million years old, was also found in France. The fossils's significance lies more in what is preserved, rather than in its age, though that too is important. Sections of the skull, beak and neck have been preserved. Having a preserved beak is noteworthy, because this is a rare find in fossils. This beak probably survived due to having been buried (and therefore protected) by fine limestone.[7] The fossilized beak is approximately 30 cm (12 in.) in length and has a joint that allows the beak to distend, levering the pouch to open wide. This ancient pelican differs little from

Joseph Smit's illustration of a Peruvian pelican (*P. thagus,* formerly *P. molinoe*).

Fossilized pelican beak.

modern species, strongly resembling the modern great white pelican, *Pelecanus onocrotalus*. It is about the same size as the smallest pelicans living today and belongs to the same genus, *Pelecanus*.[8] According to Dr Antoine Louchart of the University of Lyon, France, this pelican's beak structure tells us much about the evolution of the pelican. The fossil dates from the Oligocene epoch, 'when fish were about the same shape and size as they are today'.[9] Researchers have hypothesized that the pelican beak quickly evolved to suit the size of the fish of the time, and then ceased evolving. The lack of evolutionary change may have played a role in flight as well; if the beak had evolved to be bigger, it may have hampered the pelican's ability to fly. This means that for at least 30 million years, the beak of the pelican has remained largely unchanged.[10]

Pelican beaks are remarkable, even today. The Australian pelican, a large bird (yet medium-sized by pelican standards) with a wingspan of 2.5 m (8 ft) or so, sports the longest bill in the avian world. At 42–6 cm (16.5–18 in.) for the male, and 36–41 cm (14–16 in.) for the female, the Australian pelican is recognized as a holder of a Guinness World Record.[11]

There have been other fossil finds, including *P. intermedius* in Germany and a small pelican from South Australia, *P. tirarensis*, both from the Miocene period.[12] Pliocene fossils are more numerous, unearthed in India, Ukraine, North Carolina, Idaho and Florida. The Florida find includes fossil pelicans of the extant species *P. erythrorhynchus*.[13] Numerous subfossil pelican bones of extant species have been found in Neolithic deposits, having been killed for food.[14]

Trans-Tasman relations have often been competitive; Australians and New Zealanders have argued over matters pertaining to pavlova, cricket and rugby; even pelicans have entered the arena of one-upmanship. In New Zealand in 1930 a number of subfossil bones were found at Lake Grassmere and at four North Island sites. The skeleton found at Poukawa, one of the North Island sites, is 3,500 to 4,500 years old, the bones from the other sites are younger. It was thought to be a new (sub)species of pelican: *Pelecanus conspicillatus novaezealandiae* (Scarlett, 1966, 'New Zealand pelican'), as they appeared to be larger, weighing about 12 kg (26 lb), and with a broader pelvis than the modern Australian pelican. It was later given full species status because it seemed to have the marks of a distinct species. However, in 1998 New Zealand pride was dashed because paleozoologist Dr Trevor Worthy, associate professor at Flinders University, after reviewing new material, concluded that these pelican fossils were not indistinguishable from the Australian population.[15] The argument was that New Zealand

3068.—Head of Pelican.

Head of a pelican, from *Charles Knight's Pictorial Museum of Animated Nature: and Companion for the Zoological Gardens. Illustrated with four thousand wood engravings*, vols I–II (*c.* 1850).

never had a resident breeding population of pelicans; the New Zealand pelican is just the Australian *Pelecanus conspicillatus*, no longer species *novaezealandiae*, purely of vagrant status.[16]

As well as the subfossil remains, Australian pelicans have reached New Zealand on five occasions since 1890, when a specimen had been shot. In August 2012, owing to extensive flooding in eastern Australia, which provided ideal breeding conditions, at least fourteen pelicans made it across the Tasman Sea to New Zealand.[17] Unfortunately for these pelicans, five were shot during their stay; uneasy relations continue between Australians and New Zealanders.

Until recently New Zealand's only resident Australian pelican was Lanky, who arrived at the Wellington Zoo in 1978. He was the zoo's most long-term resident, living there for nearly forty years until he was euthanized in 2016 due to poor health. Lanky was a much-loved bird, remembered for his boldness; he would steal food from the monkeys in the enclosure next door. Lanky was not

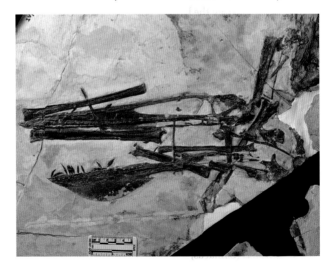

Nearly complete skeleton of the Cretaceous pterosaur *Ikrandraco avatar*.

the first long-living Australian pelican in New Zealand, though. In 1919 Percy the pelican came to live at the Wellington Zoo, where he died at the age of 62, making it into the *Guinness Book of World Records* as one of the longest-living birds – a second entry for the Australian pelican.[18]

'They look like flying dinosaurs,' said the director Judy Irving, speaking about pelicans, the subject of her documentary *Pelican Dreams*. The closest link we have to living dinosaurs is the bird. This is alluded to by mathematician Dr Ian Malcolm in the film *Jurassic Park*. As the film draws to a close, with Malcolm being flown to safety, still visibly shaken by his experiences with the dinosaurs, he peers out the window, and sees a flock of brown pelicans, 'one of the most superficially dinosaur-like of the world's 10,000 or so bird species' flying below.[19]

Looking at a pelican, one can see this ancestry; something of the dinosaur era is reflected in its features. This is not pure imagination. In Chaoyang Liaoning province, northeast China, two incomplete fossils from the Early Cretaceous era (approximately 146–100 million years ago) have been uncovered, of a pelican-like pterosaur.[20] The anatomy of its jaws suggests that this flying reptile skimmed fish from surface water and stored them in a pelican-like throat pouch. The creature, named *Ikrandraco avatar* after the pterosaur-like flying creature 'Ikran' in the sci-fi film *Avatar*, had a distinctive blade-like crest under the tip of its lower jaw, with a little hook.[21] This hook may have anchored soft tissue of an extendable pelican-like throat pouch and is a feature that had been tentatively proposed for pterosaurs, based on impressions of soft skin folds found in some fossil specimens, but had been unconfirmed until now.[22]

Returning to the present day, to the more familiar pelican: what of its family, the Pelecanidae (Rafinesque, 1815)? Its genus name is *Pelecanus* (Linnaeus, 1758). The word has its origins in the

Greek *pelekan*, which comes from the verb *pelekaō*, meaning to hew wood, and *pelekus*, an axe.[23] This has led some to propose that this word is used to describe a bird that cuts wood with its bill – a similar word, *pelekas*, means woodpecker.[24] It is a bit of a stretch to use that analogy for the pelican. Perhaps it bears the name because its huge bill resembles an axe, rather than functions as an axe, for it scoops its food, rather than cuts at it. According to a twelfth-century bestiary, the pelican derives its name from its habitat on the Nile in Egypt, which in Greek is called Canopos.[25] Old English defines the word as *pellicane* (or *pellican*), from Late Latin *pelecanus*. The spelling of *pellicane* was influenced in Middle English by the Old French *pelican*, which has remained.

Other cultures have colourful and descriptive names for the pelican. In 1868 John MacGregor (also known as 'Rob Roy'), a Scottish explorer, philanthropist and sportsman, was travelling through the Middle East. He wrote: 'The Arabs call the pelican . . . "jemel el bahr", that is, "sea-camel", which well describes its manner of carrying the head with the neck in a double arch.'[26]

Alcatraz, or *al-qādūs*, is an old Arabic word for a vessel for drawing water. It was assigned as a name for the pelican because the bird was believed to fill its huge bill with water to take to its chicks in the desert. From the sixteenth century *alcatraz* (or *alcatruz*, 'the bucket of a water wheel') became the Spanish and Portuguese word for pelican. In San Francisco Bay, the island of Alcatraz, which used to be the site of the infamous prison of the same name, was named after the Spanish word for pelican because of the large numbers of brown pelicans nesting there.

The etymology of the word for pelican in Yuwaalaraay and Gamilaraay, the indigenous language/dialect of the Gamilaraay people of northern New South Wales, Australia, is *gulayaali*, from a story about a pelican of that name who had a big net he would take out of his mouth. Later, Gulayaali used his beak to scoop up fish, like a net.[27] Gulay means 'net bag' and aali means 'two', so gulayaali means 'net-two' or 'two nets'.

Some etymology seems more myth than fact. Jorge Luis Borges writes that the name is derived from its colour: 'The Pelican of fable is smaller and its bill is accordingly shorter and sharper. Faithful to popular etymology – *pelicanus*, white-haired – the plumage of the former is white while that of the latter is yellow and sometimes green.'[28]

The American ornithologist John James Audubon first observed the white pelican in 1808, on the Ohio River, near Henderson, Kentucky. Later, he wrote: 'I have honored it with the name of my beloved country, over the mighty streams of which, may this splendid bird wander free and unmolested to the most distant times, as it has already done from the misty ages of unknown antiquity.'[29]

Pelecaniformes is an order of medium-sized and large water birds found worldwide. Traditionally the order included all birds that have feet with all four toes connected by ample webs, also known as steganopodes or totipolmate swimmers because of this webbing. There is disagreement, however, concerning which bird species belong within the order of Pelecaniformes. In the past the order was thought to consist of six families of water birds: pelicans, gannets and boobies, tropicbirds, cormorants, frigatebirds and darters. Present-day debate has historical precedent. An early member of the Royal Society, founded in 1660 for the promotion of scientific knowledge, was John Evelyn, who was fascinated by ornithology. Evelyn described the pelican as 'a fowle between a Stork and a Swan, – a melancholy Water foule'.[30] In 1758 the great cormorant was given the name *Pelecanus carbo* by Linnaeus, who called it 'a pelican with a black body'.[31] The great frigatebird, when originally described in 1789, was named *Pelecanus minor*, the lesser pelican, for it was smaller than even the smallest pelican.[32] Here we perceive (even if due to ignorance) an acknowledgement of the taxonomic debate.[33] Perhaps even then there was an understanding of sorts; that bird species thought to belong to the order of Pelecaniformes were not fixed, and that groupings were not clear-cut.

Debate and discussion continues; membership is not a closed case because of scientific advances. Evidence gleaned from skeletons (including fossils), feathers, gut morphology, carotid artery features, parasites, behaviour, DNA and egg-white proteins (which in pelicans resemble those found in the Atlantic gannet) has led some to conclude that the order Pelecaniformes may well be derived from more than one ancestral line.[34] Although ornithologists are still debating taxonomy, and which birds to include in the order of Pelecaniformes, in conclusion we note that pelicans may be quite distinct from the order to which they gave their name. What we can say with certainty, though, is that all species of pelican are placed in the single genus *Pelecanus*. When considering the range of

An 1870s lithograph of the head of a pelican, with a detail showing the foot.

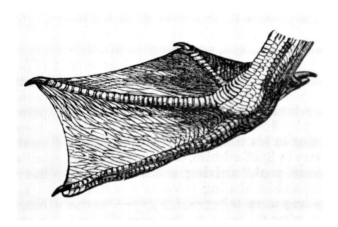

Foot of pelican, from *Charles Knight's Pictorial Museum of Animated Nature: and Companion for the Zoological Gardens. Illustrated with Four Thousand Wood Engravings*, vols I–II (c. 1850).

territories inhabited by the eight species, we find that pelicans are found in most of the world, except for Antarctica.

Pelicans, perhaps due to their comical build, have been surrounded by myth and inaccuracies for centuries. According to a French medieval bestiary of about 1450: 'Pelicans live in Egypt. There are two kinds: one kind lives on water and eats poisonous animals like crocodiles and lizards; the other kind, with a long neck and beak, makes a sound like an ass when it drinks (this kind is called the onocrotalus).'[35] According to Guillaume le Clerc, a thirteenth-century cleric, in his *Bestiaire*: 'The pelican is a wonderful bird which dwells in the region about the river Nile. The written history tells us that there are two kinds, those which dwell in the river and eat nothing but fish, and those who dwell in the desert and eat only insects and worms.'[36]

These observations are correct in two details: pelicans inhabit Egypt, and there is more than one species of pelican. The pelican family consists of four species of ground nesters and four (plus five subspecies) that are wholly or partially arboreal (tree dwellers).

Flock of great white pelicans (*P. onocrotalus*), Djoudj National Park, St Louis du Senegal, Senegal.

GROUND NESTERS:

Great white pelican: *Pelecanus onocrotalus* was named by Carl Linnaeus, the Swedish zoologist, in 1758.[37] Other common names include eastern white, European white and rosy pelican. In French it is *pélican blanc*; in German, *Rosapelikan*; and in Spanish, *pelicano común*; so the great white is known either for its white appearance or reddish hue, or for its great numbers, being 'common'. At 15 kg (33 lb) it is the heaviest of all seabirds.[38]

American white pelican: *Pelecanus erythrorhynchos* was defined in 1789 by Johann Friedrich Gmelin, a German naturalist. This pelican was discovered in Hudson Bay and New York. The scientific name means 'red-billed pelican', from the Greek *erythros*, 'red', and *rhynchos*, 'bill'. Unlike most of the other species of pelican, this pelican has only one other common name, that of 'the rough-billed pelican'.

Dalmatian pelican: *Pelecanus crispus* was named by the German ornithologist Carl Friedrich Bruch in 1832 in Dalmatia.[39] The Dalmatian pelican was at one time considered a subspecies of the spot-billed pelican. No subspecies are known to exist, but a

Nicolas Robert, *White Pelican*, 1656, gouache on vellum.

American white pelican feeding its chick, Colorado.

Pleistocene paleo-subspecies, *Pelecanus crispus palaeocrispus*, has been described, because of fossils discovered at Binagady, Azerbaijan. The Dalmatian pelican is the largest of the pelican species and its bill is the second largest of any bird.

Australian pelican: *Pelecanus conspicillatus* is the scientific name coined by Coenraad Jacob Temminck, a Dutch naturalist, who first described the pelican in 1824. Another name is the spectacled pelican, for the ring of bare skin around the eyes, a translation of *Pelecanus conspicillatus* from the Latin *conspicio* 'look [at] carefully', and *conspicillum*, 'a place to look from'. Temminck called the bird the *Pélican à lunettes*, by which it is still known in French, and in German this has been translated to *Brillenpelikan* ('spectacled pelican').[40]

ARBOREAL (PARTIALLY/WHOLLY):

Pink-backed pelican: *Pelecanus rufescens* was named by Gmelin in 1789 in West Africa. In French it is *pélican gris*; in German, *Rötelpelikan*; and in Spanish, *pelicano rosado*. These names mirror the different hues in their plumage, the pink or red, and the grey, its French name highlighting the fact that it is the most grey of all

American white pelican in breeding plumage showing horny plate on upper mandible. It is displaying with wings spread and standing on a rock in the water, Colorado.

Dalmatian pelican (*P. crispus*) taking off from water, in Gujarat, India.

pelicans and may appear quite drab next to some of the more striking species.

Spot-billed pelican: *Pelecanus philippensis* was first documented in 1789 by Gmelin, in the Philippine Islands; sadly it is now extinct in the Philippines, from where it gets the second part of its ornithological name. It is also known as the grey pelican, the rosy pelican, the spotted pelican and the Philippine pelican. The spot-billed was formerly confused with the Dalmatian, for their territories overlap.

A mass of beaks of the Australian pelican (*P. conspicillatus*), Kangaroo Island, South Australia.

Brown pelican: *Pelecanus occidentalis* (*occidentalis* is Latin for 'from the west, Western') was named by Linnaeus in 1766. It had been sighted in the West Indies. The brown pelican is also known as the American brown pelican, and the common pelican. In

Pink-backed
pelicans.

Spanish the brown pelican is known as *alcatraz*, or *alcatraz moreno* (brown or dark-skinned). In Guadeloupean Creole French, the brown pelican is known as *grand gosier*, literally 'great throat', referring to the pelican's pouch. Captain Jean Bernard Bossu, a captain in the French navy who was sent to New Orleans, wrote in the mid-1700s: 'The pelican is called the grand-gosier in Louisiana by the inhabitants because of its big pouch.'[41] Grand Gosier Island, located in Louisiana, is translated as 'Pelican Island'. The brown pelican, the smallest of the pelican species, is also the most variable, having five subspecies.

P. o. occidentalis, Linnaeus, 1766.
P. o. carolinensis, Gmelin, 1789.
P. o. californicus, Ridgway, 1884.
P. o. murphyi, Alexander Wetmore, 1945.
P. o. urinator, Wetmore, 1945.

All pelicans are classified in the genus *Pelecanus*, although the brown pelican differs from classic *Pelecanus* in a number of ways: it is marine, it plunge dives and it has dark plumage.

Robert Jacob Gordon's sketch of *Pelecanus rufescens* (pink-backed pelican), 1786, watercolour.

Spot-billed pelican
(*P. philippensis*).

Peruvian pelican: *Pelecanus thagus* was named by Juan Ignacio Molina in 1782 after being sighted in Peru and Chile.[42] It is also known as the Chilean brown pelican or the Chilean pelican. It has similar plumage to the brown pelican but it is noticeably larger, nearly twice the weight (7 kg/15 lb) and longer. The Peruvian pelican and the brown pelican are the only true marine pelican species.

Take note of the beautiful, descriptive names; from the almost comical French description of the Australian, to the hint of prison life, glimpsed in the drab of the brown. One can imagine family portraits aligned along a passage, a picture of each species, distinctive, yet similar.[43]

Brown pelican diving, Urbina Bay, Isabela Island, Galapagos.

The eight species of pelican are traditionally divided into two groups, based on plumage colour and behaviour. The first group consists of the four ground-nesters, whose adult plumage is predominantly white and who tend to fish alone (although groups do fish in lines); these are the Australian, the Dalmatian, the great white and the American white pelican, often classified as 'the big four'. They are found in four widely separated regions, though the territories of the great white and Dalmatian overlap. Their ecology and behaviour show many similarities. The other group is made up of the four grey- or brown-plumaged species which are wholly or partly arboreal, nesting in trees (pink-backed, spot-billed, brown pelican) or on sea rocks (Peruvian pelican), and

which are often colonial. They differ more from each other than 'the big four'. New research, however, using DNA sequencing of mitochondrial and nuclear genes, suggests that the relationships are quite different from the traditional groupings:

> Our data rejects the widespread notion that pelicans can be divided into white- and brown-plumaged groups. Instead, we find that, in contrast to all previous evolutionary hypotheses, the species fall into three well-supported clades: an Old World clade of the Dalmatian, Spot-billed, Pink-backed and Australian Pelicans, a New World clade of the American White, Brown and Peruvian Pelicans, and a monospecific clade consisting solely of the Great White Pelican, weakly grouped with the Old World clade.[44]

This research also hypothesizes that pelicans evolved in the Old World and then spread into the Americas. It also suggests that the preference for tree or ground nesting is based on size, rather than genetics. It will be interesting to see what other findings surface from this research.

Q: WHY WAS THE PELICAN KICKED OUT OF THE HOTEL?
A: BECAUSE HE HAD A BIG BILL.

When we consider the pelican, its bill and expandable pouch are probably the images we have in mind. Most of us are familiar with Dixon Lanier Merritt's limerick, about how much food a pelican can store in its beak and pouch.[45] In a Fijian creation tale, the pelican's pouch is a gift from a human. Long ago, a fisherman named Ratu Tatanga wove the largest fish traps of any of the fishermen. In those days, pelicans could only catch one fish at a time. Ratu Tatanga, moved by the pelican's plight, gave him a large, woven fish trap to use as a pouch; from that day on, the pelican could catch many fish in one scoop.[46]

Pelican bills are not only long, but very strong; the bill of the brown pelican is much stronger than that of the white, however, because it has to sustain the impact of water after it plunge dives from a great height. Pelicans are capable of catching fish more than three-quarters the length of their bill. The pelican is helped in this feat by having a down-curved hook at the end of its upper mandible, known as the mandibular nail, which is used to hook or kill its prey, and for preening.

Pelicans are the only bird with a pouch under their bills: 'The pouch is really a three-layered skin bag. The skin is on the outside.

'The Pelican . . .', from Leopold Joseph Fitzinger, *Bilder-Atlas zur wissenschaftlich-populären Naturgeschichte der Vögel in ihren sämmtlichen Hauptformen* (1864).

The inner layer is a mucous membrane. Between the two is a thin layer of two sets of muscles running in opposite directions.'[47] The pouch is kept in pristine condition by exercise:

> Pelicans perform strange-looking exercises to stretch and maintain their pouch in a brand of pelican yoga. They will gape, holding their mouths wide open. In another pose, they point the bill straight up to the sky, stretching the pouch. Or most evocatively, a bird will turn its pouch completely inside out by forcing it over its breast.[48]

When we think of pelicans, we usually picture them sporting huge gular pouches, resembling avian Santas with bulging sacks. This skin bag is also capable of contraction, making it difficult to see. The bony projection of the lower bill and the flexible tongue muscles form the pouch into a basket for catching fish. Once something is caught, the pelican draws its pouch to its breast. The tongue is tiny; its specialized muscles contract, tightening the pouch, forcing the fish into a swallowing position. Water is drained, the fish swallowed immediately, then predigested and regurgitated later if needed to feed young.[49] The stiff upper mandible serves as a lid for the expanded pouch. Contrary to popular belief, pelicans do not use their pouches to store (thus Merritt's limerick is incorrect) or carry live fish in water to feed their offspring; the pouch is used as a scoop net, and to catch food thrown by humans.

The pelican's tongue muscles are also used in gular fluttering, a form of evaporative cooling. Pelicans can easily overheat in the hot sun, so they shed heat by facing away from the sun and fluttering their bill pouches, which contain many blood vessels, to let body heat escape. It rapidly flutters the pouch by contracting and relaxing the muscles.

Brown pelican (*P. occidentalis*) with beak raised, displaying distinctive pouch or gullet, Florida.

The long bill and large gular pouch have evolved to enable the pelican to be a successful fish catcher, but the bill has other functions as well; it can detect fish swimming in murky water, or at night, and is a filter, excreting excess salt.

Although the bill resembles a weapon, the pelican rarely uses its bill in fighting, but during the breeding season, they may jab at one another. During copulation the male will use his bill to grab the female's neck or head and then hold her down. The pelican's long bill may play a role during the breeding season, but, unlike in many bird species, its bill is not used to aid chicks during egg hatching.

The pelican's pouch has spawned many tales, from seagulls waiting on top of the bill, ready to grab a fish (in reality, seagulls are more likely to be waiting underneath the pouch to catch fish that fall out), to the idea that pelicans use their pouches to collect rainwater. There have been sightings of pelicans opening their bills to take in rainwater, but this occasional act cannot be classed as an inherited behaviour.

It is its pouch which makes the pelican so distinctive. Would we be as interested in the pelican if it were pouchless? I don't think so. Even though many of the pelicans in bestiaries and in heraldry are depicted with smaller pouches and thinner bills, they are still, on the whole, recognizable as pelicans.

Q: WHAT FISH DO PELICANS EAT?
A: ANYTHING THAT FITS THE BILL.

The pelican is surrounded by strange tales, half-baked facts and great feats of the imagination. The pouch in particular has fascinated writers for centuries. How does it operate? Is it more like a pot than a net? This fascination may be due, in part, to its wildness; it is not a domesticated bird, hence the inability for centuries to be able to study it closely. When Aelian (Claudius Aelianus, 175–235 CE) wrote *De natura animalium* (On the Nature of Animals), a work consisting of seventeen books of natural history, he included the pelican. His writing was based almost entirely on other written sources, often Pliny the Elder's, and is therefore anecdotal in nature, but nonetheless this is an important work, for it was a reference for both medieval natural history and bestiaries, which were common during medieval times. Writing about the pelican's eating habits, Aelian suggests that a pelican's stomach is a mini furnace: 'River-dwelling pelicans are known to swallow mussels, which are then warmed up in their bellies and open up. When this happens, the pelicans puke them out and then eat the flesh.'[50]

Perhaps Aelian borrowed from Aristotle: 'Pelicans living by rivers swallow big, smooth shells. After cooking them in the pouch in front of the stomach they vomit them up so that when they are open they may pick out the flesh and eat it.'[51] In Book Ten of his *Natural History*, Pliny the Elder (first century CE) also shows an interest in the pelican's digestive system: 'Pelicans have a second stomach in their throats, in which the insatiable creatures place food, increasing their capacity; later they take the food from that stomach and pass it to the true stomach.'[52]

Dionysius of Halicarnassus (first century BCE) was another writer who had strange theories about the pelican's feeding habits. He, too, thought that the shellfish were cooked, but added:

> They do not dive completely under the water but dip their necks, which are six feet long; keeping their body above the water, they take every fish that they find. In front they have a kind of pouch into which they put all their food, refraining neither from cockles or hard mussels, but gulping everything down, shell and all. When the creatures are dead they vomit up the mouthful, eat the flesh and expel the shells.[53]

Bartholomaeus Anglicus (thirteenth century CE) mentioned the animal's dining etiquette: 'All that the pelican eateth, he plungeth in water with his foot, and when he hath so plunged it in water, he putteth it into his mouth with his own foot, as it were with an hand.'[54]

These descriptions have come, for the most part, from armchair thinkers who have relied on second- and third-hand accounts, infused with a drop or two of imagination. Although they are flawed records of biology and of ornithological exactness, they do display a sense of wonder at the magnificence of the pelican.

Observing pelicans feeding in their habitat is breathtaking, even after setting aside the fabric of legend. Watching pelicans feed has led many to marvel at their communal fishing techniques. Pamela Conley describes pelican fishing practice at first hand; while swimming, she finds herself in the middle of a catch:

> My foot hit a fish and then the ocean began to explode around us. I froze in fear. I looked up and saw a hundred pelicans swarming down . . . Some were landing as close as two feet away, and in the centre of this hysteria I realised that they were swooping in on a school of fish and we were in the middle of a feeding frenzy . . . Then understanding replaced panic and I watched in awe as pelicans scooped up fish with what appeared to be gallons of water.[55]

The American white, Australian and the great white are predominantly communal fishers: they form an arc, or semi-circle, on the water and swim together. They use their feet and wings to splash the water, which tends to drive the fish towards shallow water, where they scoop them up. Some researchers believe that the birds' fishing method is even more complex and that pelicans, perhaps with the exception of the brown, have in fact evolved to be cooperative.

Some species employ different fishing methods. The pink-backed and spot-billed usually paddle along the water, slowly swim up to the fish near the surface, then quickly scoop them up; they tend to be solitary fishers. Dalmatian pelicans, less gregarious than the great white, usually forage alone or in groups of two or three, but occasionally will fish communally.

Pelicans are also opportunists. For example, in the Terek Delta and the Volga–Caspian channel, Dalmatian pelicans fly with, or observe, thousands of cormorants. After the cormorants have dived deep, bringing fish to the surface, the pelicans scoop up the fish, sometimes even striking the cormorants in the process.[56]

'Pellicano d'America', from Saverio Manetti, *Ornithologia, methodice digesta atque iconibus aeneis ad vivum illuminatis ornate* (1767–76).

Pelicans will steal food opportunistically, but their main feeding method is communal (or cooperative) fishing, supplemented at times by scavenging, on land and behind fishing vessels, and a wee bit of 'stealing'.[57]

The brown pelican plunge dives from a height of up to 20 m (68 ft) in order to catch fish. Brown pelicans' eyes are not harmed from the impact of diving from such a height, because

CDL XXXXIX

'Pellicano', from
Saverio Manetti,
*Ornithologia,
methodice digesta
atque iconibus
aeneis ad vivum
illuminatis ornate*
(1767– 76).

their eyes are protected by clear membranes, shielding them
during diving, under water and during flight. A common myth,
though, is that brown pelicans become blind due to supposed
trauma sustained via the plunge-diving method.[58] Some brown
pelicans are blind, but this is due to the effects of pollution, not
from diving.[59] Plunge-dive accuracy is heightened by the pelican's ability to orientate itself away from the sun when diving,

Jan van Essen's brush drawing of pelicans in Artis Zoo, 1880.

which decreases surface glare.[60] When a brown pelican plunge dives, its body rotates but its head remains stable, allowing continuous sighting along its bill.[61] When the bill hits the water, the pouch contracts, so that the fish is caught between the upper and lower mandible. The lower mandible opens, catching fish and water, then the upper mandible shuts. The pelican rises to the surface, to drain water from its pouch. This method of fishing relies on speed; less than 'two seconds lapse between hitting the water and catching the fish, although it may take up to a minute to drain the pouch.'[62]

Pelicans may have long bills and spectacular gular pouches, but their external nostrils have evolved into non-functional slits, forcing them to be mouth-breathers. As one astute reviewer on a website that highlights movie mistakes pointed out, Nigel the pelican in the animated film *Finding Nemo* was drawn with nostrils – but pelicans do not have nostrils. The reviewer was partly correct; external nostrils are present, but they are non-functional.

Pelicans, despite their size, are a fairly quiet bird. Only at mealtimes or in large breeding colonies do they become vocal, grunting to express excitement or alarm, hissing if displeased. The small number of sounds is due to their anatomy: 'The syrinx is apparently unmodified; it has no special structures and no flexible tympanic membranes to produce many different sounds.'[63] The large young produce a mixture of screams, whines and yelps, which are higher pitched than adult calls, and can be heard from quite a distance. Some pelicans bill-clap loudly with heads thrown back, or bow and sway their heads to defend their territory. Pelicans rarely make a sound outside their breeding colony, though they may be noisy during cooperative feeding.

Pelican courtship rituals, as for many species, are both rich and complex. For example, in the case of the Australian pelican the female leads a group of males around the colony. During this time, the males compete for her attention, swinging their bills from side to side, hoping to catch her eye by throwing sticks into the air and catching them. Both sexes participate in the rituals of pouch rippling and bill clapping. Eventually there is one male left standing, and this lucky one is escorted by the female to a potential nest site. They may dance again, bowing their necks and pointing their bills up or down. Finally, the female sits on the nest site, the male continues to perform for her, before moving into the closing act, that of mating.[64]

When the young pelican hatches, it is quite patchy in colour, but this does not prevent parents from identifying their young from among similar-looking chicks:

A newly hatched pelican has a large bill, bulging eyes, and skin that looks like small-grained bubble plastic. The skin around the face is mottled with varying degrees of black and the colour of the eyes varies from white to dark brown. This individual variation helps the parents to recognize their chick from hundreds of others.[65]

Though perhaps not the most attractive when born, the bond between chick and parent is strong. One writer commented that 'they are useless and disagreeable domestics', but this is not so![66] Care of young is fairly equally distributed between sexes. Early care consists of close brooding and feeding, but once the chick can regulate its own temperature, most parents simply guard them and cover them at night. One Indigenous Australian observed:

When the little pelicans hatch they leave their nests and spend their time playing and learning with other baby pelicans in a kind of kindergarten. One or two adults look after lots of babies while most of the adults are out looking for food. In a lot of ways they are like the Aboriginals, sharing their campsites, and the raising of their young.[67]

Mobile young of ground-nesting pelicans do gather in kindergartens, known as crèches or pods; these groups may help with temperature control and protect the young against predators. The size varies depending on the species. Similar to the way parents recognize their chicks, young also spot their parents, and leave the pod to beg for food. 'The tendency to form pods is a clearly defined stage in development.'[68]

Let us leave the happy hearth and venture into the dark side. Obligate brood reduction, or 'the role of executioner' to give it a more dramatic name, is practised by a number of animals and birds, including several species of pelican.[69] Two or, for some pelicans, three eggs are laid, and they hatch several days apart, which is a long interval for birds. This means that by the time the second chick has hatched, the first is bigger and stronger. The newly hatched will probably be attacked by its older sibling (brood reduction is undertaken by the young, without adult participation), until, in most cases, it dies. If this is the case, we may well ask: 'What is the point of laying more than one egg?' This behaviour does not seem linked to competition for food,

Brown pelican pair in courtship at nest, Urbina Bay, Isabela Island, Galapagos.

Eggs of the great white pelican (*P. onocrotalus*).

because some pelicans have been observed attacking the second chick even before it has emerged from its egg.[70] There are several reasons why pelicans lay more than one egg, yet resort to siblicide, the primary reason being that it is a form of insurance. This practice, also known as 'twinning', was first suggested by the behavioural ecologist Dave Anderson.[71] If something is wrong with the first embryo, there is another in reserve, but if all goes well, the second egg is not needed. Pelican eggs have a long incubation period, some 32 to 35 days; if they waited until hatching time to find that their one egg was unviable, they would most likely forfeit an entire breeding season. It seems better to add a second egg than have no young that season.

The first critical test of the insurance hypothesis was conducted by the behavioural ecologist Roger Evans and his student Kevin Cash, who were both studying American white pelicans in southern Manitoba. They made a comparison study of two groups: one in which Evans and Cash removed one of two eggs shortly after they had been laid, and one in which the two eggs remained untouched by the scientists. The results were quite startling; clutches of two yielded more chicks than the nests reduced to a solitary egg, even though in the clutches of two, only one chick survived by siblicide. 'The reason was that about one in five eggs failed to hatch.'[72] Another reason for not raising both young could be about preserving the health of the parents. An additional chick may take a toll on the general body condition of the parents.[73]

Another violent aspect of the pelican, seemingly confined to ground-nesters, concerns the rough way the young may be handled before feeding. This unexplained behaviour, which does not appear to serve any useful purpose, helped fuel the legend that the pelican may kill her young, then pluck her breast to revive them with her blood.[74] Even more bizarre, though, is the frantic behaviour of the chicks. When a chick wants to be fed, it often works itself into a frenzy, thrashing about and throwing itself at its parents' feet.[75] They may bite parents, siblings, other young or even their own wings.

In the pink-backed species, feeding the young seems to be taxing; large young can take up to twenty minutes to be fed. The adult may need to rest before recommencing the feed. The fully fed young 'become grossly distended, exceptionally splitting outer skin of neck'.[76] Large young also peck their parents' bills and sometimes their feet, behaviour not recorded in other Pelecaniformes.

There is another strange behaviour that has been observed among the pink-backed young: 'A curious, common and, along with self-biting, uninterpreted behavior is to tuck its bill until the tip touches the breast, and then slowly head-shake.'[77] Convulsions usually follow, sometimes with the young pelican becoming comatose.[78] Convulsions are not confined to the pink-backed pelican; other species of pelican also convulse. The severity of the convulsions appears to be dependent upon the species. Cash and Evans noted that convulsions seem to affect 90 per cent of chicks over the age of three weeks.[79] In the American white species, chicks that had to beg longer were more likely to show aggression and/or convulse. The Australian pelican is prone to convulsions, a common occurrence after eating. Although the pelican may become comatose, this is a short-lived state, because the bird will soon make a full recovery.[80] In nests with two chicks, only the larger chick convulses. This behaviour prevents the younger from feeding, leading to its starvation.

Explanations for the convulsions are either unconvincing or incomplete. Asphyxiation, as a consequence of having its head deep in the adult's pouch for several minutes or after swallowing a large fish, has been suggested, but this does not take into account pre-feeding convulsions.[81] Cash and Evans suggest that it is an extreme form of begging, and a method of driving away other young, but this would be unnecessary in single-chick species in which adults recognize their own young.[82]

Leaving behind an analysis of some of their earthbound behaviour, let us turn our attention upwards, to consider the miracle of pelicans in flight. If we have been fortunate enough to have observed a pelican in flight, most of us would agree that they are engineering marvels. They look, and are, heavy (the great white and the Dalmatian are the heaviest flying seabirds in the world), but their pneumatic bones and air sacs mean that their skeletons weigh less than their feathers.[83] The major pectoralis muscle in the American white pelican, and presumably also in others, contains 'fast twitch' fibres in the superficial layer, which aid flapping, but in its abdomen are slow fibres, which assist in soaring flight.

Pelicans appear to be clumsy on the ground, and often have trouble becoming airborne; it can be taxing work if there is no wind. If near water, pelicans run over the water while beating their wings, pounding the water surface to get enough speed for take-off. Once airborne, though, they are majestic.

In flight, pelicans are well-organized and coordinated, often in squadron formation, flapping then gliding. In 2001 new research was undertaken to determine why birds fly in 'V' formation. French scientists from the Centre National de la Recherche Scientifique in Villiers en Bois studied great white pelicans that had been trained to fly behind a light aircraft and a motor boat for a feature film, *Le Peuple Migrateur* (Winged Migration).[84] Researchers fixed heart monitors to the birds' backs. The results were startling, demonstrating that pelicans' heart rates decrease when they fly as a group. They also change their mode of flight, gliding rather than flapping. 'They fly in formation to save energy,' said Dr Henri Weimerskirch. 'It's not because they are using the upward air stream of their neighbor, it is because they are able to glide more often.'[85] Both these factors help conserve energy, aiding the pelicans to fly further during their migratory flights. This scientific research 'suggests that formation flight evolved because it allowed birds to reduce their energy expenditure and fly further', which is advantageous for migration and for gathering food. Flying in a 'V' also allows birds to communicate with each other.[86]

Pelicans do not undertake long sea-crossings, though a few may be blown by strong winds to oceanic islands. Migration patterns among pelicans vary according to species. The pink-backed do not migrate long distances, but some regular movements are observed from the northern coast of Africa into the sub-Saharan steppe for the wet season, their movements related to water conditions. The spot-billed breeds can be found in India, Sri

Lanka and Cambodia, but in the non-breeding season can be located in Nepal, Myanmar, Thailand, Laos and Vietnam. The Australian pelican has no particular migratory pattern, instead gathering in large breeding colonies when and where conditions are favourable, such as temporarily flooded inland lakes.

Brown pelicans may remain in their breeding territory, depending on its location, and if there is ample food supply. Others move north, along both the Atlantic and Pacific coasts, returning south in the winter. Food sources may govern the times of the northern migration. If food is plentiful, the brown pelicans may leave in late summer. Some of the eastern subspecies migrate to Florida, the Caribbean coasts of Colombia and Venezuela, and the Greater Antilles. During cold winters, some Texas brown pelicans winter along the Gulf Coast of Mexico.[87]

Most of the western population of the Dalmatian pelican stays through the winter in the Mediterranean region, arriving in March and usually leaving by the end of August. It is more actively migratory in Asia, where most of the pelicans that breed in Russia fly south to spend winter in the central Middle East, Sri Lanka, Nepal and central India. Dalmatian pelicans that breed in Mongolia winter along the east coast of China.

Great white pelican migratory patterns vary. The African population is resident, but those breeding in the Palearctic region are migratory, their migration routes only partially known. Pelicans from southeast Europe and Asia migrate to Africa and Iraq, while pelicans from northeast Europe may reach China, India and Myanmar. Migratory populations are found from Eastern Europe to Kazakhstan during the breeding season, with over half breeding in the Danube Delta. The pelicans arrive in the Danube in late March or early April and depart after breeding, from September to late November. The great white pelican can fly long distances, averaging 30–40 km per hour (19–25 mph) during migration. Its

travel distance can be up to 200–300 km (124–86 mi.) per day, which is a massive feat for any species of bird.[88]

The brown pelican, from Audubon's *Birds of America* (1827–30).

Most populations of American white pelicans are migratory, though some populations found on the Texas coast and in Mexico are permanent residents. Breeders from the northern plains migrate southeast and southwest to coastal lowlands. 'Populations breeding east of the Rocky Mountains migrate south and east . . . to winter along the Gulf of Mexico. Populations breeding west of the Rockies migrate over deserts and mountains to the Pacific coast.'[89]

Wind patterns are vital for migratory birds. When flying, pelicans soar rather than flap, relying on atmospheric conditions, such as updrafts and thermals, for lift and propulsion. Recent research has been published concerning the impact of seasonal

Louis-Pierre-Théophile Dubois de Nehaut, photograph of pelicans in Brussels Zoo, c. 1855.

wind conditions on the migratory patterns of the American white pelican.[90] Scientists studied the effects of different wind factors (tailwind, turbulence, vertical updrafts) on the migratory flying strategies adopted by 24 satellite-tracked American white pelicans throughout spring and autumn in North America. They discovered that the pelicans optimized the use of available wind resources, 'flying faster and more direct routes in spring than in autumn'.[91] The pelicans fly faster in spring because they are in better physical condition, and because they are drawn to their breeding grounds. Their faster flight is enhanced by updrafts, which are present in their migratory flight path in spring but are absent in their southern autumnal flight. In autumn, when the pelicans are not at their peak, they rely heavily on tailwinds.[92]

This enables them to glide, conserving energy. These results suggest that American white pelicans, and possibly other species of pelican, use wind patterns differently in spring and in autumn. This 'plasticity of migratory flight strategies' may assist them to cope with future stresses such as climate change, which is expected to have an impact on wind patterns.[93]

The story of the pelican's origins is a tale composed of different narratives: fossils and dinosaurs, Guinness World Records and formation flight, disputes over family membership and vagrants. Rich courtship delights and the cuteness of kindergartens hide the dark thread of siblicide. Most families have a skeleton in their closets; so too do most species of pelican.

2 From Creation Myths to Concrete Behemoth

Many birds feature in religion and mythology. The pelican, though lacking the high profile accorded to other birds, such as the raven or the humble dove, has still had more than a walk-on role in religion and mythology, and graces the stage in several ancient and modern cultures.[1]

Although the pelican does not feature in many creation myths, it is prominent in one. The Seri people, a tribe who inhabited the island of Tiburon, in the Gulf of California, believed that the world was created by the 'Ancient of Pelican', who brought land to the surface after a great flood.[2] This ancient pelican, a supernatural being, was also associated with great wisdom and beautiful song.

In Indigenous Australian stories, the pelican is part of the landscape of the Dreamtime and is credited with bestowing several gifts to human beings.[3] In a myth from the Wangkamura people, Muda, a human being, transforms himself into a pelican in order to travel more extensively. During one journey, he becomes ill and needs to rest. As Muda rests, he notices the ground glowing with brilliant colours. He pecks at them, which releases a spark that fans into flames. The Wangkamura people believe the pelican gifted them two important commodities: fire and opals.[4]

In another positive myth, humans acquire the knowledge of making nets from pelicans. Goolay-Yali the pelican was the first

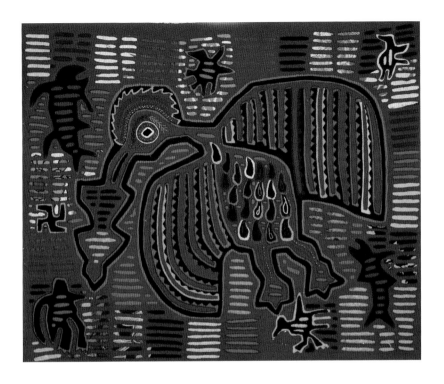

creature seen with a net, but no one knew where he kept it. Several children hid to see where Goolay-Yali's net was kept, and watched, astonished, as they saw him twist his neck and pull a large fishing net from his mouth.[5] The children could not keep this secret, so Goolay-Yali, using the design of his pouch, taught the tribe how to make nets. Although these pouches are smaller than the first net of Goolay-Yali, they are large enough to allow the tribe to bear his name, 'one having a net'.[6]

There is also, however, an Australian indigenous myth in which the pelican is cast in a less flattering role. This myth, addressing the violation of tribal law and its associated retribution, involves

Mola textile by Kuna Indian artist, depicting a pelican. From the San Blas Archipelago, Panama. Reverse applique design worn on female blouse.

Weedah, a man who absconds with two sisters. At dusk they arrive at a large river, where they meet Goolayyahlee, the pelican, who possesses a canoe. Weedah asks him to take them across the water, but Goolayyahlee says he can only take them one at a time. He takes Weedah across, then returns to take the two sisters for himself. In order to escape, the sisters throw white ash from a fire over the pelican. This myth also explains pelican feather pigmentation; since then the pelican has been white, except for where he had managed to shake off the ash, in which places black feathers remain.[7]

Another version of this myth makes reference to the uniqueness of the pelican's pouch. In this variant, Weedah takes only one female. When they arrive at a flowing river, they see an old man paddling a small bark canoe. He is described thus: 'Goolay-yali was a peculiar person, with a jaw that was half as big as his canoe.'[8] The story proceeds, with punishment meted out by the Great Spirits:

> 'Goolay-yali, you who desired a woman who craved mercy from you, the ashes with which you are covered will be the sign of your shame, now and forever.'
>
> Where he had been standing there was now a white pelican, covered in white feathers, with thin legs and a huge pouch under his beak that resembled the scuttle-like mouth of the man who had been Goolay-yali.[9]

In another story, which also serves as an explanation for the colour of Australian pelican feathers, their hue being a graphic and permanent consequence of transgressing tribal taboos, Moola the pelican covers himself with white war paint and grabs his spear as he pursues a girl. When the older pelicans see him, they are horrified and attack him, but the younger pelicans, admiring how fierce he looks, wish to be like him. They splash themselves with white war paint, ignoring their elders, who cry that pelicans are black. The young, with their white markings, remained that way, which is why pelicans now have black and white feathers.[10]

For some first peoples, superstition is attached to the killing of a pelican. Philomene Corrigal, of the Cree Nation, tells of a personal incident:

> When we caught sight of some pelicans . . . my uncle began to shoot them. Surprised by my uncle's actions, my mother turned to him and using a firm voice, she told him that he had tampered with the law of nature. Needlessly killing pelicans, she said, would bring the wind to answer this desecration.
>
> . . . By the time evening arrived . . . the wind really began to blow . . . During the night . . . my father had to leap from his bed to grab a white canvas board so he could nail it quickly against another window blown apart by the continuing fury of the wind . . . When morning arrived, the whole area adjacent to our cabin was covered with a thick blanket of grass, sticks, leaves and branches . . . That was how strong the wind blew.[11]

Several species of pelican (Dalmatian, pink-backed and the great white) spend their winters in Egypt, so it is not surprising that pelicans were assigned a role in Egyptian religion. In ancient Egypt the pelican, known as Henet, was associated with, or considered to be, the mysterious goddess Henet, depicted either as a pelican or as a human with the head of a pelican.[12] Henet was associated with death and the afterlife; according to funerary papyri, Henet was gifted with the ability to prophesy safe passage through the underworld for the recently deceased. In a passage from the

Linen textile
tapestry, Egypt,
9th–12th century.
Two pelicans on
either side of a
conventionalized
tree of life.

papyrus of Nu, from the *Book of the Dead*, the pelican is portrayed as a helper, its open beak signifying that the deceased could leave the long, narrow tomb:

> The doorwings of the heavens stand open for me.
> The doorwings of the earth stand open for me . . .
> The mouth of the pelican is opened for me.
> The pelican has caused that I shall go forth during the
>   day to every place where I wish to go.[13]

In the ancient Egyptian Pyramid texts of Unas, many of the utterances, or spells, were for protection. One, for the Protection of the Sarcophagus Chamber, has the priest proclaim: 'The Majesty of the Pelican falls in water.'[14] To single out the pelican falling into water may seem odd, but this strange reference makes sense when understood as the pelican involved in holy work, scooping up negative elements in its beak. This may have some similarities to the belief that bird nets were used for trapping evildoers in the Egyptian underworld. As well as being mentioned in texts, pelicans were also depicted on tomb walls.

Pelicans, as protectors of the dead and guardians against the danger of snakes, were viewed as holy creatures, though they were hunted as a food source, a practice that continues today, with the flesh of great whites available in Egyptian markets. In ancient Egypt, flocks of sacred pelicans (the lucky ones) were kept in solar temples and were sometimes mummified, their feathers and eggs valued as items of tribute.

Long before Egyptians mummified pelicans, however, other societies were using pelicans in the mummification process. The world's oldest mummies, dating from 6000 BCE, were created by small fishing communities known as the Chinchorro in northern Chile and southern Peru. Unlike the ancient Egyptians, they did not mummify pelicans; instead, they used pelican skin to help mummify human bodies.[15]

In ancient Peru birds were highly venerated, because they inhabited both earthly and heavenly realms, and were viewed as messengers from the sky. In 1994 Ruth Shady Solis, of the National University of San Marcos, in Lima, began excavating Caral, the oldest discovered town in South America, believed to have been a peaceful society of musicians rather than of warmongers. A total of 32 flutes made of pelican wing bones were discovered, hidden in the main temple.[16] Later, the Moche people worshipped nature and pelicans were depicted in their art. After the decline of Moche civilization, the Chimú ascended in importance. In Chimú society the majestic Peruvian pelican symbolized power, fertility and abundance. The pelican was vital, aiding the Chimú to fish successfully; fishermen followed pelicans in order to locate fish and tied them to ropes to help them catch fish. Pelicans played another key role: their rich guano was high-quality fertilizer. The accumulation was so immense that Spanish chronicler Cieza de León wrote that the guano resembled 'peaks of snow-covered mountains'.[17] The pelican also became a symbol for both religion (or belief) and politics – in Chimú art as a shaman transforming himself into a pelican, its large pouch symbolizing the statesmen who provided for the community.[18] Among the archaeological remains of Chan Chan, the former capital of the Chimú Empire, ten walled citadels have been unearthed, decorated with raised clay friezes that include pelican motifs.[19]

The importance of the pelican in ancient Peru did not carry over into Judaism and the Hebrew Scriptures, where the pelican is assigned a less flattering role. The Hebrew word for 'pelican' is *ka'ath*, which means to 'disgorge' or 'vomit', a reference to the way pelicans provide food for their young.[20] Pelicans regurgitate

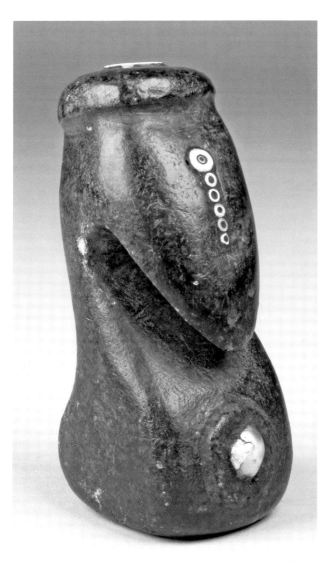

Chumash stone pelican with shell inlay, 16th–17th century.

predigested fish to their chicks, but so do other birds; it is unclear why the pelican species is singled out.

In the King James version of the Bible (KJV) the pelican is listed in Leviticus 11:18 and Deuteronomy 14:17 as an 'unclean' bird, but in other versions a different species is named, usually 'the desert owl'. In the footnotes of several translations scholars state that correctly identifying several of the birds, including the pelican, is impossible. We see this confusion in Isaiah 34:11 (also in the KJV) and Zephaniah 2:14, where *ka'ath* is usually translated as 'owl'. The pelican's classification as being ritually unclean or impure is still a matter of contention within traditional Jewish texts, and that conundrum continues in later translations.

In the King James Bible the pelican, as well as being classed as unclean, assumes a melancholy role: 'I am like a pelican of the

An earthenware
double-whistle
decorated
with pelicans,
800 1350, Peru.

wilderness' (Psalm 102:6). These words have puzzled modern translators, because the pelican is a water bird and wilderness usually refers to the desert. This can come down to our personal image of what constitutes 'wilderness' (or 'wildness'); oceans or lakes can, at times, be desolate places. As Michael Bright comments, 'it is . . . possible that the seventeenth-century scripture writers were good observers. When the conditions are right, the pink-backed pelican may abandon its aquatic habitat and head for the desert. Here it feeds not on fish but on locust swarms.'[21]

Other naturalists have noted that sometimes when a pelican's pouch is full it will fly to a nearby desert area to digest its food: 'Often it will sit on its breast there for days in a most melancholy attitude, giving the Psalmist's image even further strength.'[22] It is worth noting that, depending on their habitat, some pelicans favour lakes that are situated in deserts, such as Walker Lake and the Great Salt Lake.

An amazing reversal, though, is that the pelican, deemed 'unclean' (depending on which translation you read), was elevated to a prominent and noble role within Christianity and is frequently depicted in Christian art and literature, church architecture, furnishings, carvings, misericords, chalices and in the ecclesiastical heraldry of bishops and other notable religious figures.[23] In medieval and Baroque art, the image of a pelican often adorned altars and tabernacle doors; sometimes even the tabernacle itself was fashioned in the form of a pelican.

Processional cross with pelicans, Italy, 16th century.

Gerrit Willem Dijsselhof (1866?–1924), undated etching of King Nimrod as a pelican.

How did the pelican, 'unclean' in Jewish texts, become a symbol of virtue? The positive status of the pelican in Christianity is due to a powerful mix of theology, medieval mythology and legend, and confused biology. From the middle of the second century, the pelican began a long association with Christianity, mainly due to the *Physiologus*, a popular early Christian work penned by an anonymous author from Alexandria.[24] This work, which consists of a collection of legends about various animals and birds, with allegorical interpretation, includes an ancient belief about the pelican's behaviour with its young: 'The little pelicans strike their parents, and the parents, striking back, kill them. But on the third

Anonymous engraving of a cross with a pelican on top, print after Aegidius Sadeler (1666).

day the mother pelican, repenting of the deed, strikes and opens
her side and pours blood over her dead young. In this way they are
revivified and made well.'[25]

According to Walter Harter, writer and bird enthusiast, there
is a grain of truth in this: 'Male pelicans, although they guard the
nest, and even bring food for the young, sometimes – either accidentally or because of provocation – kill them.'[26] It is interesting
to note that that this legend crosses cultures; in an Indian fable,
the pelican also treats her young so roughly that she kills them.
Later, having regretted her act, she revives them with her blood.[27]
The pelican, instead of being condemned as a murderer, is elevated in stature, rewarded for what became a saving act, and is
now seen as a Christ-like figure. It may be difficult to see how the
mother pelican, killing her young, can be aligned with Christ.
Apparently, fault is with the chicks (us): we have struck our parent
(God) with our sin, but Christ's blood revives and saves us.

Epiphanius of Salamis (310–403), bishop of Salamis, drew on
this legend in his work *Panarion*. Epiphanius' version differs in that
the mother pelican is said to kill her young by her excessive kisses,

A Pelican feeding
her young, from
a French bestiary
of *c.* 1270.

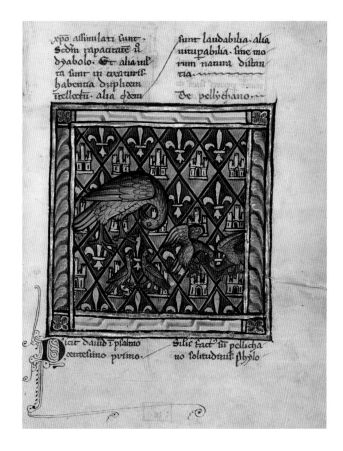

which is a gentler interpretation. The legend from *Physiologus*
influenced, and appeared in, later Latin bestiaries during the
medieval period. In medieval art and literature, the symbol of
the parental slaughter of their chicks lost favour, overtaken by the
more positive image of the mother pelican feeding her young with
her own blood, even if that led to her own demise.[28] By the time
that Honorius of Autun (1080–*c.* 1154) wrote the *Speculum Ecclesiae*

(Mirror of the Church), the pelican had become a powerful symbol. Honorius and others believed that the pelican waited three days in order to restore her chicks to life with her blood, to mirror the Christian belief in Jesus' resurrection from the dead. Honorius stated that the pelican 'revives them (the chicks) at the end of three days . . . even as on the third day God raised his Son'.[29] There is also a Eucharistic element; the pelican feeds her young with her blood, embodying Jesus' words at the Last Supper: 'This is my blood . . . shed for many.' In some versions of the legend the mother pelican, after reviving her chicks with her blood, dies, making the ultimate sacrifice. It is not difficult to see how the pelican transformed into a metaphor for Christ.

There are variants. In one version, the chicks are killed by a serpent and then revived by the mother's blood. The fourth-century scholar St Jerome, in his commentary on Psalm 102, also blamed the serpent for the death of the chicks. Jerome wrote that the serpent was the Devil, the offspring, mankind trapped in sin, and the pelican was Christ, who shed his blood to save mankind. Leonardo da Vinci's fable, which includes the snake and the death of the mother pelican, concludes on a sentimental, mawkish note:

At the sight of such carnage she began to weep, and her lament was so despairing that all the inhabitants of the forest heard it and grieved.

'What meaning has my life without you?' said the poor bird, gazing on her dead children. 'I want to die with you!'

And she began to peck with her beak at her breast, just over the heart. The blood gushed out of the wound and poured over the baby birds killed by the snake.

But all at once the dying pelican gave a start. Her warm blood had restored life to her children. Her love had revived them. And then, happy at last, she herself died.[30]

Another version of the legend describes the pelican as having the greatest love among the creatures for its young, so much so that it feeds them with its own blood when other food is scarce. In any case, the main point is that people believed that the pelican was willing to wound itself in order to save its offspring.

These legends gave birth to the image of a pelican with wings spread feeding its young with drops of its own blood. This traditional composition of the pelican and its young is known as 'the pelican in her piety, vulning herself', from the Latin *vulno*, 'to wound'. This depiction began to spread widely, the pelican

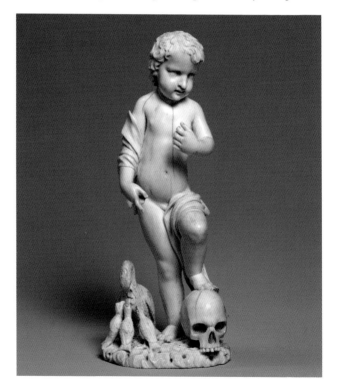

Flemish(?) ivory statue of the Christ Child with pelicans, 18th century.

becoming a frequently used motif in Christian literature and art. The pelican sometimes appears in paintings of the Crucifixion, representing Christ, one such work being *The Passion of Christ and the Pelican with her Young* by Hieronymus Bosch. Many portrayals of the pelican in bestiaries and, later, in heraldry appear more like a hawk or an eagle, drawn by people who had not seen them: 'Monks were arm-chair naturalists, and their art-work shows how they wrestled with descriptions of animals provided by explorers of far-away places . . . What could be stranger than a bird carrying a bag beneath its beak – the pelican?'[31]

If we put to one side the version of the legend that tells of the pelican killing its young and instead reflect on the variant that demonstrates sacrificial love, the ultimate in good parenting, then we may well conclude that the legend possesses noble qualities. But what are the origins of this legend? Does it have any basis in fact? Fact is embedded within confused observation. During times of famine, it was thought that the pelican plucked its breast and fed her young with her own blood and, as a result, saved their lives (a variant has the mother pelican dying). In the wild, when a pelican returns to the nest and opens her pouch so that the young can eat, her breast may appear to be red (the Dalmatian pelican has a blood-red pouch early on in breeding), but here we are seeing the inside of the bill, not her breast. Some pelicans have a reddish tinge on their breast plumage, and this rosy hue, combined with the young sticking their heads into the pouch to feed, could give the appearance of breast-pecking and blood release. When the pelican appears to be stabbing itself, it is probably pressing its bill onto its chest in order to fully empty its pouch. The red tip at the end of its beak could also be mistaken for droplets of blood. Another possible reason for the confusion is the activity of preening. Maureen Lambourne, an ornithological art expert and keen observer of bird behaviour, makes a similar observation: 'The sight of the bird preening its front with its large beak could have given the impression that it was plucking at its feathers and drawing blood from its breast.'[32]

All these factors not only contributed to the legend that the pelican drew blood from its own breast, but created the impression that it was, indeed, factual. By the seventeenth century this notion had acquired scientific status. In 1673 the Royal Society published an account of pelicans in Upper Egypt with the observation that 'Some will have a Scar in the Breast, from a wound of her own making there, to feed (as is reported) her young with her own bloud, an action which ordinarily suggests devout fancies.'[33] In 1869 the question of whether the pelican feeds its young with its own blood was considered in zoological circles, once again removing it from the gallery of myth and legend and placing it firmly in the scientific wing. Mr Bartlett, superintendent of the gardens of the Zoological Society, proposed the idea at the conclusion of a paper he delivered about the similarities in feeding behaviours among the male horn-bill and the flamingo (according to Bartlett, both species eject a red fluid into their mates' gullets): '"Have we here," says Mr Bartlett, "an explanation of the old story of the pelican feeding its young with its own blood? I think we have . . . and it may be that, in the translation, the habit of one bird has been transformed to the other . . . I have yet to learn if the same power may not exist in the pelicans."'[34]

During the medieval period the image of the pelican became associated with personal piety. This concept was popularized through the mystical writings of St Gertrude of Helfta (1256–1302), who, in a vision, saw Christ in this form:

After her Communion, as she recollected herself interiorly, our Lord appeared to her under the form of a pelican as it is usually represented, piercing its heart with its beak.

Marveling at this, she said: 'My Lord, what wouldst Thou teach me by this vision?' 'I wish,' replied our Lord, 'that you would consider the excess of love which obliges Me to present you with such a gift . . . Consider also, that even as the blood which comes from the heart of the pelican gives life to its little ones, so also the soul whom I nourish with the Divine Food, which I present to it, receives a life which will never end.'[35]

During the fifteenth and sixteenth centuries artists began to depict the pelican as an allegorical figure, representing Charity and the virtue of self-sacrifice. In a French Book of Hours from about 1430, an illustration shows Charity personified: on her head there is a nest containing a pelican and her young.

Christian symbolism of the pelican was also drawn on by writers. In Dante's *Paradiso*, the third and last section of his *Divine Comedy*, Christ is viewed as humanity's Pelican: 'This one is he who lay upon the breast Of our true Pelican [*nostro Pellicano*]' (xxv, line 113). John Lyly, in *Euphues* (1606), wrote: 'Pelicane who striketh blood out of its owne bodye to do others good.'[36] One of the most powerful poems aligning the pelican with Christ came from the pen of John Skelton (1460?–1529). In *Armorie of Birds*, he wrote:

Then sayd the Pellycan:
When my Byrdts be slayne
With my bloude I them revive.
Scripture doth record
The same dyed our Lord
And rose from deth to lyve.[37]

The Welsh poet Rees Prichard's long poem *Christ is All in All* states the pelican's association with Christ very clearly:

A 1559 engraving after Pieter Bruegel showing almsgiving; 'Charity' is seen with a pelican on her head.

Christ is the Pelican, so kindly good,
That heals his young-ones with his flowing blood,
And brings them back to light and life again,
When they were by the wily serpent slain.
Christ is the Pelican, so kindly good,
That heals his brethren with his heart's dear blood,
And brings them safely back to life again,
When they, thro' sin, had been by Satan slain.[38]

The Sacred Heart, with pelican incorporated, in an undated Portuguese print.

An engraving of a pelican in her piety was included on the bottom of the title page of the first edition of the King James Bible (1611). The addition of the pelican is noteworthy because, as a concession to Puritan tastes, this edition had only four

78

Eisaqui o Tabernaculo de Deos com os homens. Apoc. 2l.

IHS

Para que tenhaõ vida. S. Joaõ Cap. 1o

S.Smo CORACAÕ DE JESUS.
O Em.mo Snr Card. Patriarca Conc. 100 dias de Indulg. a Cada pessoa q
rezar hum Padre Ne Av. Mediante de qual q destas es lampdas

em Caza de Fran.co Me. no fim da Rua do Passeio   Lça illustrations, the others being biblical figures. The humble pelican, placed below these, is significant, for the image functions as a stark symbol of Christ's sacrifice.

Pelican religious imagery can be found embedded within several hymns. For example, the sixth verse of *Adoro te devote*, written by St Thomas Aquinas to be sung after Communion, includes the words 'Pie Pellicane, Jesu Domine, / Me immundum munda tuo sanguine' (O Loving Pelican, O Jesu Lord! / Unclean am I, but cleanse me in Thy Blood). [39]

The pelican has not always been an accepted symbol in the Church, though. A hymn by poet John Bennett which narrates

Pelican
embroidery.

the legend of the pelican was not included in the *Hymnal of the Episcopal Church* (*H* 82) because it was thought too graphic:

As from her bloodied heart the pelican
gives life to given life and dies to save,
So Christ upon the cross, So God in Man
took iron through his flesh to close the grave.[40]

It was removed at the last minute by the general Convention of the Episcopal Church, the reason being that it went against the grain of tradition; yet the symbolism of the pelican, linked to Christ's self-sacrificing nature and act, is part of Church tradition: 'but never before had Episcopalians been asked to sing about it. Some delegates at the convention argued against the hymn because the image cannot be proven; others said the vivid language was too painful; others simply felt that the writing was too weird to serve as a hymn.'[41]

The exclusion of this hymn was a strange decision; perhaps some delegates felt that by addressing other creatures, it ran the risk of removing the Church from the human realm. Although not included in the 1982 hymnal, the pelican symbol continues to be utilized by church groups, including a lay Catholic group in Louisiana known as The Brown Pelican Society of Louisiana, and The Pelican Foundation, established by the Anglican Diocese of Canberra and Goulburn, in Australia.[42]

Another myth about the pelican, which links it again with the religious life, is that it consumes the smallest amount of food necessary to maintain life, thereby symbolizing those who fast. In the Aberdeen Bestiary, which dates from around 1200, the hunger of the pelican is to be imitated: 'the life of a hermit is modelled on the pelican, in that he lives on bread but does not seek to fill his stomach.'[43] For many though, the pelican represents the opposite;

due to its large pouch, the pelican is more likely to symbolize gluttony than asceticism.

The pelican has been included in the titles of several religious books, harking back to references in the King James Bible. 'The pelican in the wilderness', quoted from verse 6 of Psalm 102, is the title of Ivan Clutterbuck's autobiography – the notion of being on his own in inhospitable territory perhaps alluding to his work in the Anglo-Catholic movement.[44] The quote has been used to title other books too. *A Pelican of the Wilderness*, the auto-biography of Robert W. Griggs, a parish minister, tells of his being hospitalized with major depression and his subsequent recovery.[45] Isabel Colegate's *A Pelican in the Wilderness: Hermits, Solitaries and Recluses* is a series of meditations on solitude, the opening quote being: 'A man that Studies Happiness must sit alone like a Sparrow upon the Hous Top and like a Pelican in the Wilderness.'[46] Unlike Clutterbuck and Griggs, both of whom felt alone, Colegate addresses those who seek solitude. Most species of pelican, though, are communal, and do not tend to be on their own; imagery from Psalm 102 does not equal avian facts. Most pelicans require others, in addition to its mate, to provide optimal conditions for successful breeding. Pelicans are colonial, with some colonies numbering thousands of pairs. Pelicans tend to be gregarious, and crowd close to each other, even if their breeding ground is spacious. Sometimes breeding colonies contain a number of different pelican species, especially the Dalmatian and the great white pelicans. Sometimes other water birds fulfil that role, happy to be in the same breeding space: 'and the presence of other nesting water birds, particularly ciconiiforms, has been known to stimulate reproduction in the wild, as also in captivity'. When breeding in captivity (the first pelican to do so was in Rotterdam, in 1872) a group of at least twenty other great whites is required 'for social stimulation and breeding synchronization'.[47]

·SCTA·CATERINA·

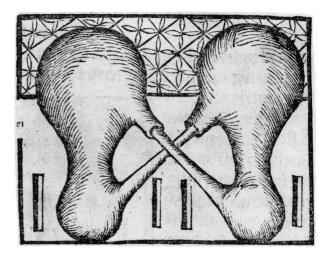

Alchemical apparatus from the 16th century: the double pelican.

This is the main reason for the absence of offspring from the pelicans which reside in St James's Park, London. In 1995 this issue was raised in the House of Lords. Lord Stoddart, concerned about the arrival of two new pelicans, asked: 'Is it kind to import those pelicans and so deny them a normal life with a mate, including the patter of tiny webbed feet?' Lord Inglewood reassured Lord Stoddart that pelicans cannot produce fertile eggs unless they are part of a larger group. 'As for the nature of the community in which pelicans live, it is similar to that experienced in monasteries and nunneries,' he said.[48]

In the 1700s the Freemasons adopted the pelican image from one of their forebears, the Rosicrucians. The pelican in her piety became a prominent symbol of the Eighteenth, or Rose Croix, Degree of the Ancient and Accepted Scottish Rite in Freemasonry, because the pelican symbolized Christ and resurrection.[49] At the base of the Rose Croix (a rose-entwined cross, the jewel of the Eighteenth Degree), there is a pelican feeding its young. There are seven pelican chicks, seven being an esoteric number in the Eighteenth Degree.[50] According to Conor Moran, a Freemason, up until the eighteenth century the pelican in her piety was mostly portrayed piercing itself on the right side of the breast:

> Freemasons began portraying the pelican piercing its left side as a symbol of self-sacrifice required of its members. Many artists and craftsmen were unaware of the significance of the right side of the breast and so began the act of using the pelican of Freemasons in Christian art and churches.[51]

In the proto-scientific tradition of alchemy, birds were highly valued because they lived in different realms and were gifted with the ability to transform themselves. The pelican symbolized the Philosopher's Stone that dissolved (died) in order to allow gold to rise from lead; the destruction of the old, to make way for the new.[52] The legend of the pelican piercing its breast indicated that personal sacrifice was needed in order to follow the path of alchemy.[53]

The pelican is also utilized in animal guidance work, usually symbolizing selflessness. This trait has been observed in the pelicans' nesting behaviour and communal fishing methods.[54] Mention is made of its pouch. 'Are you trying to store what shouldn't be stored? Are you not using or digesting what you have?'[55] The brown pelican's plunge-diving method is drawn on as well; after it dives, it recovers: 'The pelican holds the knowledge of how to rise above life's trials.'[56]

The pelican, fuelled by the legend that highlights its attributes of altruism, self-sacrifice and compassion, was adopted as a symbol by several blood companies. The logo of the Irish Blood Transfusion Service is that of a pelican, and for many years its headquarters were located at Pelican House in Dublin. The Northern Ireland Blood Transfusion Service in Belfast also uses a pelican logo. A pelican in its piety is depicted on a lavishly illustrated poster for the former Central Laboratory of the Netherlands Red Cross Blood Transfusion Service (CLB). Sanquin, its successor organization, has as its emblem a stylized pelican with a drop of blood on its chest. On their website they include the legend, which specifies the species of pelican (the Dalmatian), a 'fact' not mentioned in the early legend. The pelican's self-sacrifice is held up as an example to model: 'Just as the mother pelican so selflessly saved the lives of her offspring, so do generous donations of blood save the lives of patients.'[57]

Pelicanus

Tab. 62

Pelican blood donor, *c.* 1650.

CHRISTIANISSIM
ET INVICTISSIMO.
)REAS LAVRENT

André du Laurens, roundel of a pelican above the head of King Henry IV, from an illustration in his *Historia anatomica humani*.

The importance of the pelican, at least as a symbol, is not confined to religion, spirituality or to charitable acts. The pelican has also influenced governments and has forged a long association with royalty, particularly with queens. Pelican links with royalty first arose as the result of the resolution of a messy governmental dispute. In 1392 Richard II quarrelled with the London civic authorities and moved the royal court to York. Reconciliation finally took place, largely due to Queen Anne's efforts, and the royal

court returned to London. At Christmas, the grateful Londoners presented her with a pelican as a token of their thanks. The symbolism highlighted Anne's maternal love for her subjects as well as her sacrifice for them. There are conflicting reports regarding the form this gift took. The *Westminster Chronicle* recorded that Queen Anne had been presented with a bird brooch, possibly in the form of a pelican, during the season of Epiphany, in 1393.[58]

Nicholas Hilliard, portrait of Queen Elizabeth I wearing a pelican pendant, *c.* 1575, oil on wood.

Detail of the pelican pendant from Hilliard's portrait.

Spanish(?) jewelled gold pendant with a pelican inset, 17th century.

Elsewhere, the *Westminster Chronicle* wrote that the city repaid Queen Anne at Christmas 1392 with a gift described as 'a large and marvelous bird, with a very wide gullet', in other words a pelican.[59]

Later, the pelican endeared itself to Queen Elizabeth I and became one of her favourite symbols. In the 'Pelican' portrait of her, attributed to Nicholas Hilliard, Elizabeth is depicted wearing a pelican pendant. The pelican pendant represents self-sacrifice and charity, and by wearing it Elizabeth is stating that she is the mother of the nation, sacrificing all for her subjects. Pelicans were included in other works of art, too. In a red-chalk drawing by Federico Zuccarao, a pelican is perched on a column immediately behind the queen, and a similar portrait, attributed to Crispjin van de Passe the Elder, was used as an engraving for a book frontispiece. Engravings of Elizabeth I in book frontispieces were common, as this was a way of acknowledging the queen's presence in the home. For those who knew of the pelican legend, the pelican was not merely a physical bird, as its image symbolized the personal qualities and virtues of the queen and her reign.

Queen Anne was not the only member of royalty to have received a pelican as a gift. Pelicans were first introduced to St

Frederick William Bond, photo of keepers feeding the pelicans in London Zoo, October 1915.

James's Park, London, in 1664, a gift from the Russian Ambassador to King Charles II. These pelicans, great whites from Astrakhan, in the Volga Delta, were described by writer and diarist John Evelyn as looking 'very melancholy'.[60]

Pelicans were also subjects of the Royal Menagerie. In 1822 when, for the first time in the menagerie's six-hundred-year history, it came under the professional supervision of a zoologist, improvement in conditions led to creatures, including pelicans, being more prepared to breed: 'The pelicans from Hungary nested, for what was probably the first time in captivity.'[61]

Over the years, more pelicans have been added to the several in St James's Park. At one time, these pelicans played a role in Cold War politics. During the 1960s, a U.S. Ambassador visited the Foreign Secretary. They met in a room which overlooked St James's Park, its lake and pelicans in full view. The U.S. Ambassador was briefed on the history of the pelicans, including that the Russians had been giving pelicans as gifts since 1664. Concerned about this action by the Soviets, the U.S. Ambassador arranged for several pelicans from the United States to be given to the Royal Park. When they arrived, these pelicans ignored their Russian feathered friends, and seemed rather miserable. 'The U.S. Embassy suspected the Soviet Embassy of harming the American pelicans – which the Russians denied – and relations between the embassies became glacial.' Ornithologists were consulted. The problem was not due to diplomatic intrigue or foul play; rather, it turned out to be the species of pelican. These were brown pelicans (*Pelecanus occidentalis*), pelicans which favour saltwater. The freshwater lake in St James's Park was an unsuitable habitat for them, and they were soon relocated to a zoo. The U.S. Ambassador proceeded to present a number of American white pelicans to the Royal Park, a species content with freshwater, and they soon settled down with their Russian neighbours. This prevented a possible diplomatic incident. This story is denied by some, including staff of the Royal Parks: 'We're aware of this story, but it's now impossible to confirm the detail,' cautions a Royal Parks spokesman.[62]

The gifts from Russia continue. In 1977 the Russian government, through the UK/USSR Committee on the Environment, gave two great whites, named Astra and Khan, after the Astrakhan nature reserve where they had been reared, following in the tradition of the first pelicans. In 1995 four new pelicans came from Prague Zoo (two for St James's Park and two for London Zoo). The following year a new pelican, later named Gargi, joined St James's Park, not as a gift from a foreign government or dignitary, but as a stray, having landed in a garden in Essex.

In 2013 the decision was made to acquire several new pelicans to join the ageing ones. In collaboration with The Royal Parks, Prague Zoo and The Tiffany & Co. Foundation, three new great white pelicans, two females and one male, joined the descendants of the original Russian flock. The programme took eighteen months to bring them over, acclimatize them and then release them. The new ones include Tiffany and Isla, named after the daughter of the wildlife officer who has cared for the pelicans for over 35 years. All is well, with the exception of the occasional eating of a pigeon, an act which was promptly filmed and downloaded onto YouTube.

Pelicans also feature in the art and science of heraldry. The 'Pelican in her Piety' signifies filial devotion and a willingness to sacrifice one's life. In medieval heraldry the pelican is usually portrayed with the head and body of an eagle, hawk or phoenix, wings elevated, pecking at its breast. Like the artists who produced the medieval bestiaries, whose work they would have consulted, early heraldic artists had to work, for the most part, from hearsay and legend. The pelican is always shown plucking its breast. One striking change to note, though, is that in heraldry the pelican is usually male: 'This is suggested by Guillim [the source of most modern British heraldry], who notes that "the Aegyptian priests used the Pellican for a Hieroglyphick to expresse the four duties of a Father toward his children."'[63]

The pelican is also prominent in ecclesiastical heraldry and carvings. On the badge of the Guild of Corpus Christi, the pelican appears above a nest containing its young. The pelican is depicted on the coat of arms of the Corpus Christi Colleges at both Oxford and Cambridge, perhaps a reference to the Colleges' name, 'corpus christi' meaning 'the body of Christ'. At Corpus Christi College, Oxford, the pelican is depicted on the college's crested tie, the college flag and on the top of the Pelican Sundial; at Corpus Christi, Cambridge, the image graces the college's crockery.

The pelican legend of self-sacrifice is incorporated in the arms of a number of English and European families and by clergy. Some of the mottoes for the pelican have included *Ut vitam habeant* (That they may have life), *Immemor ipse sui* (Unmindful herself of herself), *Mortuos vivificat* (Makes the dead live) and *Nec sibi parcit* (Nor spares herself).[64]

Flag of Louisiana, 1861 (left), and Louisiana's new state flag (right), featuring a more angular pelican tearing its breast to feed its young, plus the addition of three drops of blood, was unveiled in 2010.

American Civil War envelope. The shape behind the shield may be a pelican.

In the United States, Louisiana is known as 'the Pelican State' because brown pelicans were plentiful there until the latter part of the twentieth century. Pelicans have been a symbol of Louisiana since the early 1800s, appearing on state buckles, buttons and flags. The use of the pelican on the flag has a long and rich history, pelicans depicted on troop flags as early as the Mexican–American War of 1846–8. The earliest state flag, from December 1860, featured a red field with a white star, a pelican feeding its young painted within the star. Although there have been a number of different flags used over the years bearing pelicans, some blue, some red (depending on the military companies and organizations), it took until 1 July 1912 for the Louisiana State Legislature to officially declare the pelican device as part of the state flag, displaying the state bird, the eastern brown pelican, in white and gold on a field of blue. As in heraldry, this pelican did not have a large pouched beak.

The 1912 flag would probably have remained unchanged, but for the zeal of Louisiana high school student D. Joseph Louviere. In 2005, while researching a social studies project, Louviere noticed the lack of uniformity in the pelican device on Louisiana flags and seals. Some had drops of blood on the pelican's breast (the droplets varying in size and number) and others had none. Historically, there are three pelican chicks in the design, so a drop of blood is required for each, the three drops of blood on the pelican's breast representing the state's willingness to sacrifice itself for its citizens. Despite this rich evidence, state law did not specify the required number of drops. In 2006 'three drops of blood' was mandated by law, for both the flag and seal.[65] In November 2010 a new flag was adopted, depicting a more realistic brown pelican, with a yellowish forehead, tearing its breast, with the required three drops of blood – all due to a high school student's project.

As well as being the state bird of Louisiana, in several countries the pelican has gained prominence as a national bird. The brown pelican is the national bird of the Turks and Caicos Islands, of Saint Kitts and Nevis (the pelican image supports the shield on their coat of arms), and in Romania the great white pelican is the national bird.

Pelicans have also represented countries by appearing on stamps and marked coinage. Albania, home to two species of pelican, the Dalmatian and the great white, featured the Dalmatian pelican on its 1 lek coin. On coinage from the British Virgin Islands, the brown pelican was included on the 50 cent piece. Pelican illustrations have featured on stamps from many countries, sometimes as purely decorative, at other times assuming an educational role, promoting altruistic blood donation (and displaying the motif of the pelican in its piety, plucking its breast), or highlighting conservation issues, particularly in the United States and in Romania.

Images of pelicans have not been confined to flags, coinage and philatelic products. Illustrations of pelicans have been used to advertise a wide assortment of products, from apples, coffee and beer, to cigarettes, tobacco and even Elastoplast. The connections

Pelican House in Old Town, Warsaw.

pelicans have to these items can be quite vague, especially when the object is a type of fruit or vegetable. Perhaps the cigarette/tobacco association is because the pouch resembles an ash tray? Or that pelicans resemble a group of laughing larrikins? The beer may conjure up images of relaxing on a beach or beside a river. It may be harder to find a link for King Pelican coffee. The answer is provided on their website; the coffee was named King Pelican after its founder Ken Fine drank 'the perfect cup of coffee' in a café in Seattle called The Pelican Coffee Bar, named so by its proprietor, who used to live in Florida and remembered the pelicans there. The Elastoplast association is purely commercial; Pelican Pat was a cartoon character used in material promoting the merits of sticking plaster as a first-aid tool.[66] The picture of Pelican Pat, with Elastoplast on his knees, is anatomically incorrect; the illustrator has drawn the strips on what is actually the bird's ankle.

Apart from the benefits of Elastoplast, the pelican has been incorporated in the medical field as an acronym for a UK cancer group. The Pelican Cancer Foundation was established in 1993 to promote development in bowel, bladder, prostate and liver

Detail of pelican
planter, outside an
apartment block
in Miami, Florida,
photographed in
c. 1930s.

cancer treatment. The name 'pelican' describes the organs of
focus: 'curing and improving quality of life for patients with
'pelvic area (bowel, colorectal, bladder, prostate) and secondary
('liver') cancer'.[67] The acronym 'pelican' is also used in some
countries, not for medical reasons but for road safety. 'Pelican'

has become the accepted term for a type of road crossing used by pedestrians, assimilated from the descriptive 'pedestrian light-controlled crossing'.[68]

With their large bodies and distinctive bills, pelicans cannot help but be the subject of several mammoth endeavours, some bordering on the kitsch. A mega sculpture, billed as 'the world's largest pelican' (joining the tradition of 'the world's largest merino' in Goulburn, Australia, and 'the world's largest buffalo' in Jamestown, North Dakota), was constructed at, fittingly, Pelican Rapids, Minnesota. There, the world's largest pelican statue, named Pelican Pete, also known as 'The Mother of All Pelicans' or 'The Pelican Rapids Pelican', stands at the base of the Mill Pond Dam on the Pelican River. Built in 1957 out of plaster and concrete, it is 4.7 m (15.4 ft) high. Some propose that the pelican statue is the reason for successful fishing; when the fish see the large pelican's eye as they pass in the stream, they become disoriented, and are easily caught.[69] A number of smaller pelicans, decorated in different designs, are located outside local businesses, as part of a 2007 community art project.

The Big Pelican is not an idea confined to Pelican Rapids. In Noosa, Australia, there is another Big Pelican. The 3-m (10-ft) structure, known as 'Percy' or 'Pelican Pete', built in 1977 as a float for the Festival of the Waters Parade, has moveable parts that can be operated from the inside, using levers, pulleys and cable. At the time, the pelican was the emblem of the Noosa Council, and the statue featured on postcards and tourist material.[70]

In 2001 a public art project took place in Seabrook, Texas. The Pelican Path Project consisted of thirty decorated fibreglass pelican statues placed around the region, in honour of the return of brown pelicans (and some American whites) to the bird sanctuary in the area. Seabrook is located in one of the largest migratory paths in North America, so the project was a way to educate the community about the significance of the region for bird life. In 2004 in Pensacola, Florida, a similar project, known as Pelicans in Paradise, was undertaken, this time as a means for raising money for a literacy programme.

A costumed anthropomorphic pelican was the first mascot in the history of World Expositions. In 1984 New Orleans hosted the Louisiana World Exposition, with Seymore D. Fair (or Seymore de Faire), a pelican wearing a blue tuxedo jacket, large top hat, spats and white gloves, as the official mascot. He travelled the world, promoting the New Orleans World Fair. During this time Seymore became quite the celebrity, mixing with a number of well-known personalities, including George W. Bush and Billy Joel. Seymore even made an appearance on the popular TV show *Saturday Night Live*. For the first year, he did not have a name, known simply as Mr Pelican, until a naming contest was instituted. Out of the 18,000 suggestions, the winning entry was a derivative of the local Yat dialect. Seymore also helped promote a number of educational and community causes, with a focus on anti-drug programmes. When the Exposition concluded in November 1984, Seymore's new home was the Smithsonian. Seymore was also inducted into the New Orleans Historic Collection and the Louisiana State Archives. In 1986 the New Orleans mayor issued a proclamation declaring 24 January as 'Seymore D' Fair' day. Did Seymore raise the profile of the avian pelican? Perhaps choosing the pelican as the mascot raised some awareness of the presence and plight of brown pelicans in the region, or perhaps its twee elements led to little benefit, if any, for the actual bird and its struggle in the state of Louisiana.

In another part of the world, a real pelican became a hit with its citizens. In 1985, in Tokiwa Park, Ube City, Japan's first artificially incubated pelican was born. This great white pelican, named Katta-kun (or Cutta-kun), became a celebrity; he flew to kindergartens and schools in the vicinity, dancing as the children sang to him. He became so popular that authorities decided to curb his activities by clipping his wings. Katta-kun became the mascot of Ube City; for many years cartoon images of him graced the sides of Ube buses and drink machines, postcards and maps. In 1994 an anime film was made about him, as a fundraiser for the city. The beginning of the film reflects what happened in the life of the real Katta-kun. In the animation, Katta is also artificially incubated, and is popular with the local children, but then the story changes dramatically. In the anime one of Katta's friends is a boy whose parents are stationed in the Middle East. The boy, secure on Katta's back, is flown to the Middle East, to make sure his parents are safe. Katta appears to be endowed with super-powers, including possessing the ability to fire death rays from his beak. The largest investor in this quite ridiculous film was the tourist board of Yamaguchi Prefecture.[71] The real Katta-kun died in 2008, aged 23.

The anatomy and flying ability of the pelican have made their way into the name and design of a number of aircraft, and items carried on board planes, including a u.s. Navy guided bomb. The saaf Shackleton 1716, call-sign 'Pelican 16', was the first Avro Shackleton built for the South African Air Force. On 12 July 1994, while flying to the uk, Pelican 16 had engine failure, leading to an emergency landing in the Western Sahara. Although found, Pelican 16 is, for the records, still 'missing-in-action' and remains at the crash site – perhaps the only true 'pelican in the wilderness'.[72]

The Ultravia Pelican, a series of high-wing, single-engine tractor ultralight aircraft designed by Jean Rene Lepage, harnessed the fundamentals of pelican anatomy: a large body, yet a lightweight design. Variants include Le pelican, Super Pelican, Pelican Club, Pelican Sport and Pelican aula 600. The pelican's huge body and flying finesse were taken into account in the design of the Boeing

Pelican ULTRA, intended as a large-capacity transport craft for military or civilian use, and, if it had been built, would have been the biggest plane in the history of aviation. Its main mode was to fly low over the ocean, like its namesake, exploiting the aerodynamic benefits of ground effect that reduces drag and fuel burn. If seas were rough, the Boeing Pelican, imitating the feathered variety, could easily climb to a higher altitude and continue its flight. In 2002, after several promising press releases, information about the making of this ambitious new aircraft ceased. The most likely reason was that the design encountered serious technical problems, but this did give rise to conspiracy theories.[73]

Paper planes have also incorporated pelican anatomy in their designs. The 'Pelican', a name given to a paper aeroplane based on an H. Riley Watkins design, has a large wing area and can float and remain airborne for long periods of time, like its namesake. Another paper plane, called the White Pelican, was a collaborative year-long project between NASA engineers and eighteen high school students. It made it into the 1993 *Guinness World Records* for being the world's largest paper plane, gliding in at 35 m (114 ft). It now hangs amid jet fighters and space capsules at the Virginia Air and Space Center. 'The Pelican' has even made it into the XBox world, the name of a fictional modern-day U.S. War Plane in the game *Halo* and into the world of SLARF (Second Life Avatar Review File).

Pelican design graces at least one university's curriculum. In an engineering course at Boston University, the pelican's airborne ability, plus its unique pouch design, are factors considered for future inventions:

The pelican's versatility with the water, if replicated, can be used to explore sea and coastal life. The gular pouch is particularly interesting because it could be used to retrieve different specimen for study . . . Creating a mechanical version of the Great White Pelican may indeed be difficult . . . However, the amount of good that would come out of it certainly makes the strange bird a worthwhile challenge to recreate.[74]

The pelican has also been linked with vessels crossing oceans, sometimes associated with famous ones. When Francis Drake left Plymouth in 1577, his ship was known as the *Pelican*. Mid-voyage, Drake renamed his ship the *Golden Hind*, in honour of his patron Sir Christopher Hatton, whose crest was a golden hind.[75] Even though a coin was struck of the *Pelican*, we tend to associate Drake with the *Golden Hind*. Would we have considered the bird differently if seen alongside Drake? Would Queen Elizabeth I have preferred the original name, the pelican being one of her symbols?

In society urban myths abound, and the pelican has had a part to play in one disturbing tale. In 1984 a newspaper story told of a pelican scooping up a chihuahua who had been eating some of the pelican's fish. A crowd gathered as the pelican flew off with the little dog in its beak. The article reached the desk of Bill Scott, an Australian folklorist, who immediately suspected it was an urban myth. Over the next few years other reports were brought to Scott's attention; localities and the breed of dog changed, but the skeleton of the reports remained the same, with one report making the ABC news.[76] A fax from a vet, however, appears to validate the story:

In 1981 I examined a small, bruised and dazed chihuahua in Forster . . . The story was that the owner had been walking along the lakefront with the dog and passed a group of fishermen feeding pelicans . . . One of the pelicans swooped on the dog, mistaking it for a fish and struggled to take off with the dog scrambling around in its beak.[77]

This urban myth has filtered throughout Australia. John Ayliffe, known as 'the Pelican Man of Kangaroo Island', alluded to knowledge of the myth in an interview he gave about the eating habits of pelicans: 'They'll mug hard-working cormorants, throttling them until they drop their catch. They'll scavenge . . . pigeons that stray too close.' He has heard of a pelican swallowing a chihuahua, which doesn't disturb him: '"Useless bloody dogs, chihuahuas," he said.'[78] There may be the occasional incident of a pelican scooping up a small dog, but this is certainly not part of general pelican behaviour. The downside, of course, is the bad press that ensues for this magnificent bird.

The popularity of the pelican stretches from a rich history of association with religion and royalty, to firm links with advertising, product placement and the kitsch. From the virtue of self-sacrifice, to the consumption of chihuahuas and pigeons, pelicans have been present in our history, and remain embedded within our culture. Has their alignment, or image association with products

Pelicans entering Noah's Ark, Byzantine mosaic, 13th century, St Mark's Basilica, Venice.

Low Countries brass plate with motif of pelican piercing her breast to feed her young, 15th century.

or events, aided or hindered them? Does a pelican mascot prompt questions regarding the conservation of endangered pelicans? Does the pelican in its piety have to sacrifice itself, to die, in the natural world because of pollution? How ironic it would be if the legend of the pelican were to become reality.

SPOONY

3 Gracing the Page and Screen

Pelicans, probably due to their strange anatomy, with their large throat pouches, squat bodies and long bills, are well and truly ensconced in art and literature. They are one of the most frequently used animal caricatures from the wild and have attracted the attention of writers and artists through the centuries. Pelicans are found in verses of poetry, in Lear (Shakespeare and Edward), and loom large in picture books. They burst forth in myriad ways in film and in the realm of kitsch. Pelicans have been portrayed, at times correctly, at other times fancifully, in a variety of styles. Within most of these endeavours, artists have managed to convey something of the pelican's essence, be its majesty in flight, its self-sacrifice within the confines of legend, its comical behaviour or merely its breathtaking beauty.

As well as being portrayed in a range of ways, as characters in books and film, pelicans grace the publishing world in another form, located in book shops (especially the second-hand book variety), libraries and in many homes. Pairing up with another bird, Penguin, led to a successful partnership for over fifty years. When Penguin Books expanded into non-fiction in 1937 it did so under a new imprint, Pelican Books. Pelican, with the bird as a white silhouette on a powder blue spine and cover, became shorthand for non-fiction. Allen Lane, founder of Penguin Books, came up with the name when he heard someone who wanted to buy a

Misfitz-Spoony: a card from the 20th-century Misfitz Card Game pack.

Penguin book at the King's Cross Station bookstall incorrectly asking for 'one of those Pelican books'. Lane's goal was to introduce non-fiction to the working person, at a low cost; Pelican Books was later praised for functioning as an 'informal university for '50s Britons'.[1] After closing in 1984, Pelican Books made a

Mark Catesby, *The Wood Pelican* (*Tantalus loculator*), c. 1731–43, hand-coloured engraving.

A pencil drawing of pelicans by Ceźanne, 1870s.

triumphant return as a digital imprint in May 2014, with the same aim as before.

Lane was not the first publisher to use the pelican in its title; the Pelican Publishing Company, a medium-sized publishing house in Louisiana, was established in 1926, its history tied to well-known names such as William Faulkner. This, too, was not the first time the pelican had been incorporated in print form; in the sixteenth-century eminent English printer Richard Jugge used the pelican as his printer's device.[2]

A contemporary tale, in the manner of a modern fairy story, has been penned to displace the noble status of the stork as the bird who delivers babies. It makes a lot of sense: instead of a stork bringing a baby, wrapped in a cloth suspended from its beak, the pelican's large bill and extendable pouch is a safer and more comfortable means of transportation for newborns. In the tale, there was a storm and the pelican stepped in to help deliver a baby: 'Since that night Pelicans took over the baby delivering process . . . one day soon . . . the pelican will get the full credit in this mythological tale.'[3]

Melchior
d'Hondecoeter,
*A Pelican and Other
Birds ('The Floating
Feather')*, 1680,
oil on canvas.

The pelican is not a bird one usually associates with the emotion of love; doves, yes, nightingales, perhaps, but pelicans? The above story links pelicans with the love of a new babe, but they have also been utilized to mirror romantic love, though reflected through the dark side of legend. Richard de Fournival (1201–1260) used the form of a bestiary and animal similes to describe romantic love. His work was more than a commentary on love; it was a challenge to the Platonic idealism of love. The legend of the pelican is drawn on to address the agonies of unrequited love:

So it is too with THE PELICAN . . . For the pelican is a bird which loves its babies wondrously. It loves them to the point that it will play with them very willingly . . . becomes angry and kills them. And when he has killed them, he repents. Then he lifts his wing and pierces his side with his beak, and he sprinkles the babies he killed with the

Gerrit Willem
Dijsselhof
(1866?–1924),
undated ink
sketch of pelicans.

blood that he draws from his side. In this way he brings them back to life . . .

And so, fair, very sweet beloved . . . newness of the acquaintance had made me as it were your chicken . . . and you esteemed me so little in comparison with you that my words displeased you. Thus you have killed me with the sort of death that pertains to Love. But if you were willing

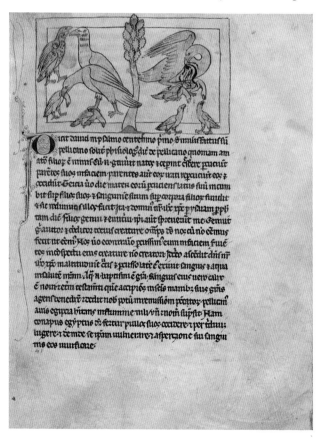

Pelicans in a mid-13th-century English illuminated manuscript.

to open your sweet side so that you sprinkled me with your good will and gave me the fair, sweet, desired heart that lies within your side, you would have resuscitated me.[4]

The legend of the pelican in its piety has influenced many, including Shakespeare. A variant of the legend, in which the pelican chicks strike back at their parents, meant that the pelican, conventionally associated with piety and self-sacrifice, could represent adolescent wickedness and rebellion instead. In Shakespeare's tragedy *King Lear*, the negative connotations are referred to in Act III, Scene 4:

'Nothing could have subdued nature
To such a lowness but his unkind daughters.
Is it the fashion that discarded fathers
Should have thus little mercy on their flesh?
Judicious punishment! 'Twas this flesh begot
Those pelican daughters.'

Lear likens himself to a mother pelican, sacrificing himself, giving up his blood/life, so that his greedy daughters can thrive. Shakespeare, rather than praising the kindness of the parent pelican in giving its blood to its chicks, considers the destructive appetite of its young.

In *Richard II*, the legend of the pelican is alluded to. Gaunt accuses Richard of consuming his family's blood: 'That blood already, like a pelican, / Hast thou tapp'd out and drunkenly caroused' (Act II, Scene 1). In *Hamlet*, though, Shakespeare reverts to the positive symbolism, having Laertes sing the praises of the life-giving factors of the pelican: 'To his good friends thus wide I'll ope my arms; / and like the kind life-rendering pelican, / repast them with my blood' (Act IV, Scene 5).

One person who devoted much of his creative output, artistic and literary, to the subject of the pelican was Edward Lear. From nonsense verse to beautifully executed illustrations to grace John Gould's *The Birds of Europe*, Lear gained a deep understanding of and appreciation for the pelican. His execution of large birds, including the pelican, is breathtaking, a reminder that Lear was an ornithological illustrator and painter before becoming better known as a writer of nonsense verse:

> They are certainly among the most remarkable bird drawings ever made, and it is evident that Lear endowed them with some measure of his own whimsy and intelligence, his energetic curiosity, his self-conscious clumsiness and his unselfconscious charm. No bird has ever . . . consorted in such grotesque conjugal bliss as his Dalmatian Pelican.[5]

On his travels abroad, Lear recorded his observations of pelicans, mainly because pelicans appealed to his sense of humour.[6] In 1849 Lear was amused to see 'many thousands of Pelicans all together' at Avlóna, on the west coast of Albania:

> I resolved to examine these mysterious white stones forthwith, when, lo! On my near approach, one and all put forth legs, long necks and great wings, and 'stood confessed' so many great pelicans, which, with croakings expressive of great disgust at such ill-timed interruptions, rose up into the air in a body of five or six hundred, and soared slowly away to the cliffs north of the gulf.[7]

On a journey to Egypt, he was overwhelmed by the variety of bird life. In his diary entry for 9 January 1867, he wrote: 'O queer community of birds! On a long sand pit are . . . 8 pelicans

– careless foolish'.[8] He travelled to Egypt more than once, and the pelicans along the Nile were the inspiration for his poem 'The Pelican Chorus':

King and Queen of the Pelicans we;
No other Birds so grand we see!
None but we have feet like fins!
With lovely leathery throats and chins!
Ploffskin, Pluffskin, Pelican jee!
We think no Birds so happy as we!
Plumpskin, Ploshkin, Pelican jill!
We think so then, and we thought so still!
We live on the Nile. The Nile we love.
By night we sleep on the cliffs above;
By day we fish, and at eve we stand

THE PELICAN CHORUS.

On long bare islands of yellow sand.
Later, as different birds gather:
Wing to wing we dance around, –
Stamping our feet with a flumpy sound, –
Opening our mouths as Pelicans ought,
And this is the song we nightly snort; –
Ploffskin, Pluffskin, Pelican jee![9]

This nonsense poem was written around 1867 but not published until ten years later, set to music in *Laughable Lyrics* in 1877. One biographer wrote that these birds, rather than bringing a lightness to the author, elicit a sad response from the now ageing Lear: 'By then of course he was a middle-aged gentleman, composing an after-dinner song, and the collection of birds who come to the pelicans' daughter's feast will convey a certain heaviness, a sense of sadness.'[10]

For Lear, happiness was being surrounded by people he loved, but this state often eluded him. Lear's poetry conveyed his lone-liness, and his grief for the past, which would never return. This precious past, often symbolized by golden shores or beaches, appears in many of his poems and songs, including 'The Pelican Chorus'. His writings were an escape from his shyness, his per-ceived ugliness, clumsiness and shame associated with his epilepsy. Did he choose particular birds to reflect something of his own physicality? 'The birds that Lear claimed as his own were indeed outstanding, and outstanding among them were the large ugly ones. Lear saw into the soul of the raucous raven, the oracular owl and the preposterous pelican.'[11] Did Lear identify with birds? 'There were times, he declared, when he felt that he was becoming a bird and he didn't particularly mind. The fantasy was with him for the rest of his life.'[12] In Lear's self-portraits there is something of the avian about him, resembling the pelican in his plumpness and stance.

As in the majority of Lear's verse, 'The Pelican Chorus' has an unhappy ending. The King of the Cranes flies off with the Pelican's daughter, having won her 'with a Crocodile's egg and a large fish-tart'; the Pelicans remain, sitting on the sand, grieving for the daughter they do not expect to see again. The repeated chorus: 'We think no Birds so happy as we!' drips with irony.

Even when Lear considers the after-life, the pelican makes an appearance. In a letter to Chichester Fortescue, a British liberal politician, Lear wonders what heaven would be like. Of course, Lear's musings were fanciful, with an unfortunate outcome for the pelican in the heavenly realm: 'In the next eggzi stens you and I and My lady may be able to sit for placid hours under a lotus tree a eating of ice creams and pelican pie, with our feet in a hazure coloured stream and with the birds and beasts of Paradise a sporting around us.'[13]

In John Ciardi's poem 'The Reason for the Pelican', the pelican, following in the tradition of Edward Lear, lends itself to the comical. In Ciardi's poem, the poet struggles to discern the reason for the pelican's large beak. After exhausting a number of ludicrous possibilities, including the beak being a place to house his wife, or used to bale out a boat, or being an extra item of footwear, Ciardi concludes that it is a 'splendid beak', even if we are still left in the dark concerning the reason for its 'splendid' size.[14] Pelicans seem to encourage the writing of the absurd. Robert Desnos (1900–1945), a French surrealist, wrote the poem 'Le Pélican', which addresses what one does with pelican eggs.[15] The poem is still a part of the French curriculum, and is used to teach French as a second language.[16] Other poets have penned odes to the pelican. In Judith Wright's poem 'Pelicans', after a verse dealing with the dark nature of certain creatures, she considers the pelican, not through the lens of judgement, or by dint of comedy, but rather by virtue of a more precious attribute, that of compassion. Wright describes this attribute via a modern example that would resonate with most humans, the pelican being as 'kind as an ambulance driver'.[17]

Other writers have utilized the imagery of the pelican. In *The Seaboard Parish* (1868), George MacDonald's description of someone hungry for information is contrasted with those who were full of news, like a pelican's gular pouch: 'In the evening, when I went into her room again, having been out in my parish all day, I began to unload my budget of small events. Indeed, we all came in like pelicans with stuffed pouches to empty them in her room, as if she had been the only young one we had, and we must cram her with news.'[18]

Poster advertising the 1936 WPA Federal Writers' Project book *Birds of the World*.

In James Joyce's *Ulysses*, a reference is made to Leopold Bloom, the fictional protagonist of the book: 'He says this, a censor of morals, a very pelican in his piety'.[19] Bloom is portrayed as a sacrificial victim, reflected by the change of the pelican's gender. This is not a first, though; in heraldry, the pelican in its piety is considered to be male rather than female.

Pelicans, due in part to their comical looks, are popular characters in children's books. They appear in Hans Christian

Roelandt Savery,
The Garden of Eden, 1618,
oil on panel.

Andersen's fairy tale *The Garden of Paradise*, but in a serious role, accompanying the prince on his journey to the Garden of Paradise: 'Pelicans flew away in rows like fluttering ribbons, and bore him company to the boundary of the garden.'[20] In Roald Dahl's 1985 book *The Giraffe and the Pelly and Me*, the pelican also works, but in a less salient job, as a window cleaner. Dahl exploits the pelican's anatomy, demonstrating the usefulness of its gular pouch.

In children's literature, the pelican is often credited with wisdom, perhaps alluding to its ancient history. In *The Great Wungle Bungle Aerial Expedition* the pelican helps save a wallaby, deciphers a message balloon and builds a satellite![21] In another picture book, *The Best Beak in Boonaroo Bay*, the pelican's wisdom again comes to the fore; he prevents the birds from fighting over who possesses the best beak.[22] The story has an interesting subtext; the reader realizes that the pelican judge has the best beak of all the species, though the pelican never draws attention to this fact. The pelican's wisdom is infused with humility, a quality that enables peace to reign within the bird community.

In Graeme Base's *My Grandma Lived in Gooligulch*, the narrator's grandmother decides to travel to the seaside and leaves . . . by pelican:

> The eagle wouldn't take them,
> And the coot was far too weak,
> But a pelican consented,
> So they climbed into its beak.[23]

Twenty-four years after its publication, the book was made into a musical. The flyer for the stage show depicts the grandmother in the pelican's pouch. Base wrote: 'I fear this was a blatant case of false advertising since we didn't have the budget to make a giant pelican for Grandma to fly off in, instead having to make do with a make-believe biplane. I wonder if anyone noticed?'[24]

The best-known book about a pelican, though, is *Storm Boy*, an Australian novel by Colin Thiele first published in 1963.[25] It gained a new following when it was released in a picture book version in 1974, with atmospheric illustrations by Robert Ingpen. Ingpen admits pelicans are a difficult subject to draw: 'they're not easy to draw, pelicans, but I can speak pelican.'[26] In 1976 it became an award-winning film. Thiele's story is of a young boy growing up in a beach shack with his reclusive father, Hide-Away. Significant issues are tackled within the narrative, including friendship, grief and loss, and change. The boy, nicknamed 'Storm Boy', has an affinity with nature: 'Sometimes he wished he'd been born an ibis or a pelican.'[27]

One day, at the wildlife sanctuary, several adult pelicans are killed. Storm Boy, who had witnessed the cruelty, finds three tiny pelicans still alive in their broken nest. He carefully picks them up and takes them home. The three pelicans survive and are later released. One of them, named Mr Percival, chooses to return to

A porcelain and
enamel plate,
c. 1745, with
pelican design.

Australian pelican
with beak open,
2013. There have
been sightings of
pelicans opening
their bills in
order to take in
rainwater as it falls.

Storm Boy. During the duck-shooting season, Mr Percival harasses
the shooters, flying in circles, warning the ducks; eventually he is
shot by an irate hunter. The final pages are poignant, the writing
exquisite, portraying grief among those reluctant to show it. The
waves of grief are not confined to the death of the pelican; they
also encompass the loss of Hide-Away's wife, Storm Boy's mother.

Mr Percival's death heralds change, indeed transition, in the lives of both Hide-Away and Storm Boy:

> All day long Storm Boy held Mr Percival in his arms. In front of the rough iron stove where long ago he had first nursed the little bruised pelican into life, he now sat motion-less and silent . . . Now and then he smoothed the feathers where they were matted and stuck together, or straight-ened the useless wing. But in his heart he knew what was happening. Mr Percival's breathing was shallow and quick, his body and neck were drooping, and for long stretches at a time his eyes were shut. Then, suddenly, they would snap open again, clear and bright, and he would snacker his beak softly in a kind of sad, weak smile, before dozing off again.
>
> 'Mr Percival,' Storm Boy whispered, 'you're the best, best friend I ever had.'
>
> Hide-Away didn't light the lantern. Instead, the three of them stayed on in front of the little fireplace – Hide-Away, Storm Boy, and Mr Percival – while darkness filled the humpy and the stars came out as clear and pure as ice.
>
> And at nine o'clock Mr Percival died.
>
> Only then did Hide-Away move. He got up softly, and, gently, very gently, took Mr Percival from Storm Boy. And Storm Boy gave him up.[28]

After they have buried him under the look-out post, Storm Boy says that he is ready to go away to school:

> And everything lives on in their hearts . . . And always, above them, in their mind's eye, they can see the shape of two big wings in the storm-clouds and the flying scud

– two wings of white with trailing black edges – spread
across the sky

    For birds like Mr Percival do not really die.[29]

In 1976, when Thiele's book was transferred to screen, the
pelicans in the film were reared from hatchlings. They were trained
by a dolphin trainer, who spent nine months with them. They
viewed him as a surrogate mother, perhaps even imprinting on
him.[30] In 2009, when the pelican who starred in the film as Mr
Percival died, his death made the news. He had been living at the
Adelaide Zoo since the late 1980s and was in his mid-thirties
at the time of his death. 'Bird keeper Brett Backhouse said Mr
Percival had a good life, with similarities to the film: "I think the
ending of it, you know, saying that Mr Percival never dies, he lives
on in other younger pelicans would be a sort of nice memento to
him . . . And Mr Percival does have a few children around the
place so yeah, he'll live on."'[31]

    It took over a decade for the book to be made into a film, and
35 years before it was adapted for the stage.[32] For the stage production, sophisticated puppetry was used, with flying birds on
sticks and birds on wheels, the latter created in a similar fashion
to the goose in the stage production of *War Horse*.

    Robert Ingpen has been involved in other writing projects
which include pelicans. In *The Voyage of the Poppykettle* he drew
on Peruvian mythology. For the Peruvians, the brown pelican was
closely associated with El Niño, either in a mighty role, symbolizing El Niño's cosmic power, or in a more down-to-earth manner,
as El Niño's spy, keeping a close watch on human beings.[33] When
humans obey the laws of nature, all goes well, but when they break
the laws, there are devastating consequences, including fierce
storms, severe temperature changes and a lack of fish. In *The Voyage
of the Poppykettle*, the brown pelican befriends a family of gnomes

Chinese porcelain saucer with pelican and fish, second half of the 19th century.

who steal a sacred vessel and sail off to seek a new home. They watch out for the Sea god, El Niño:

Fishermen said that El Niño roamed these waters with his serpent boat and pelican slaves. For centuries he had been hunting the Great Ocean Fish. He never caught it, and this made him so angry that, every few years, he changed the ocean currents and warmed the seawater. This made the anchovy fish swim to other places and the pelicans and the fishermen went hungry.[34]

In Ingpen's second book in the series, *The Unchosen Land*, the Peruvians (gnomes) explore Australia. At the beginning of their exploration, they come across some strange Australian animals and birds. 'The exception was the pelican. He was similar to their friend Brown Pelican, who had helped them to find the *Poppykettle* four years before. He was a different colour but he spoke the same language.'[35]

In the third book of the series, *The Poppykettle Papers*, this time with text by Michael Lawrence, Ingpen wrote a preface to explain the genesis of *The Poppykettle*. 'The vision of El Niño chasing fish in his serpent boat with pelican slaves on page 2 appears on a 400 year old clay pot made by Inca craftsmen of Peru. I saw the pot in an archaeological museum in Lima, and copied the picture while working in Peru in the early 1970s – so began the tale of *The Poppykettle Papers*.'[36]

In *The Poppykettle Papers*, the first night they are at sea, the Peruvians are visited by El Niño: 'It was him all right . . . coming towards us with his entourage of pelican slaves and flying fish.'[37] When they arrive in the new land, they meet a white pelican, who is portrayed as a guardian or protector of the land; perhaps a reference, or nod, to pelican wisdom: 'The first acquaintance Aloof made on his arrival was the white pelican, who seems to make it his business to personally greet all newcomers in order to ascertain whether their influence on his country is likely to be good or bad.'[38]

Pelican narratives are not confined to the printed page, they also wing themselves into celluloid, gliding through the currents

A parcel-gilt and bronze fireplace surround from the 1930s, with pelican design.

of film and animation. Some of the movies are well known, such as *Storm Boy* and *The Pelican Brief*, but the majority, including cartoons, reach a smaller audience. *Nicostratos, Le Pélican* (2011), a French–Greek production based on the book by Eric Boisset, is similar to *Storm Boy*, telling the story of a young boy living an isolated life on a Greek island with his father, after the death of his mother. The boy is unable to connect emotionally with his father until, as in *Storm Boy*, a young pelican enters his life and becomes the catalyst for healing. The movie was filmed on the islands of Milos and Sifnos, using pelicans from the film *Le Peuple migrateur*. Eight white pelicans from the Parc des Oiseaux of Villars-les-Dombes were transferred to Greece, and on the island of Sifnos a village was created for the training of the pelicans.

Pelicans have made appearances in more commercial films. In the highly successful animation *Finding Nemo* (Disney/Pixar, 2003) one of the characters is a pelican named Nigel. He saves

White pelican in Mykonos, 2011.

Dory and Marlin after they had been ingested by his pelican friend, Gerald, and Gerald then saves them from being eaten by seagulls. There is a problem with Pixar's depiction of Nigel; he is portrayed as a brown pelican, *Pelecanus occidentalis*. The film is set in Australia, where brown pelicans do not reside. Australian pelicans, *Pelecanus conspicillatus*, are quite different in appearance and habits. A number of viewers have noted this online and made comments concerning Pixar's insufficient research.[39]

Pelicans have featured in other feats of animation, some of questionable quality. One which did not make the grade was *The Adventures of Paddy the Pelican* (1950), created by Sam Singer. This animation was of such poor quality, accompanied by unimaginative storylines, that it aired for only one month. Its glaring flaws, however, meant that it did acquire some fame from appearing on *Jerry Beck's Worst Cartoons Ever*.[40] *Camp Lazlo*, an American animated television series created by Joe Murray, premiered on the Cartoon Network in 2005 and was more successful, having a three-year run. It features a large cast of anthropomorphic animal characters, including a pelican named Mort.

Apart from *Finding Nemo*, pelicans have played roles in other Walt Disney productions. Disney's short animation *The Pelican and the Snipe* (1944) has, as its setting, the Second World War. Monte (a pelican) and his friend Viddy (who is a snipe, a type of wading bird) live on top of a lighthouse in Uruguay. Monte sleepwalks (sleep-flies?) and Viddy stays awake to keep him safe. It is a gentle film about friendship, set against the horror of war. The lighthouse association continues in another Disney animation, a Donald Duck cartoon called *Lighthouse Keeping* (1946). Donald, the lighthouse keeper, annoys a sleeping pelican by aiming the lighthouse light on its nest. This upsets the pelican, leading to a duel between the two birds, who keep switching the light on and off.

Wartime usually heralds a labour shortage, because people have enlisted or been conscripted and are away serving their country. This issue is addressed in another cartoon set during war years, *Keep 'Em Growing* (1943). On one human-free farm, the animals keep the farm going, attending to the chores, rather than making a break for freedom. The pelican (hardly a farm animal) uses his pouch to help the squirrels strain the hot food.

Pelicans are popular in other cultures too, as has been noted regarding *Nicostratos, Le Pélican. Rasmus Klump*, a Danish comic strip for young children, launched in 1951 by Carla and Vilhelm Hansen, has a pelican named Pelle as one of its main characters. The comic strip was extremely successful, selling more than 30 million copies worldwide, its popularity expanding into restaurants, theme park rides and licensed merchandise. In 2016 Germany's Studio Soi, publishing giant Egmont and German pubcaster ZDF teamed up for the new CGI nursery school series *Klump*, updating the 1950s product.

It is fitting to conclude with the popular book *The Pelican Brief* by John Grisham (1992) and its corresponding film (1993). The novel, a legal suspense thriller, is about a large oil company that wants to drill on federally protected land. The Louisiana marshland is protected because it is the habitat of the brown pelican, an endangered species in that region. An environmental group opposes the drilling plans and initiates court proceedings. The federal court, seen as the guardian of the brown pelican, supports its right to survive. The work, though fictitious, highlights environmental concerns. Brown pelicans were once on the endangered list; oil spills and habitat destruction continue to decrease their numbers. Although the book and film deal with corruption and power, the plight of the pelican is very much its subtext.

Through the words of poets and writers, and sliced within the art and celluloid of film and animation, pelicans have been

The Pelican, an engraving from an 1820s edition of *Natural History of Birds*.

PELICAN.

appropriated in order to convey important lessons about grief, loss, change and conservation, or drawn on for comic relief. Some of these topics are grounded in reality or originate deep within the ancient legend of the pelican, but with others, it is harder to detect a link. It would be fair to suggest that the character of the pelican, with its unique anatomy which lends itself to comical depiction in word and pen, will not disappear from picture books, but the knowledge and use of the legend may, however, slowly slip away.

An early 17th-century Dutch silver figure-top spoon with ornamental pelican.

4 The End of the Golden Age

What do you call a pessimistic pelican?
A pelican't

.

The relationship between pelicans and humans is, for the most part, an unhappy one, marked by habitat destruction, pollutants and hunting. This change, characterized by a deterioration of idyllic conditions for the pelican, was noted back in 1802 by explorer Matthew Flinders, after having visited several islands off the coast of South Australia. Flinders suggested that the arrival of Europeans in this area would herald the passing of what he romantically termed 'the golden age of the pelicans'. He knew there would be ramifications for his discoveries, and, of course, he was correct, for his voyage of exploration paved the way for the increasingly rapid colonization of *Terra Australis*. Flinders's prophetic voice thunders down the ages:

> Flocks of the old birds were sitting upon the beaches of the lagoon, and it appeared that the islands were their breeding places; not only so, but from the number of skeletons and bones there scattered, it should seem that they had for ages been selected for the closing scene of their existence. Certainly none more likely to be free from disturbance of every kind could have been chosen, than these islets in a hidden lagoon of an uninhabited island, situate [sic] upon an unknown coast near the antipodes of Europe; nor can anything be more consonant to the feelings, if pelicans have

any, than quietly to resign their breath, whilst surrounded by their progeny, and in the same spot where they first drew it. Alas, for the pelicans! Their golden age is past; but it has much exceeded in duration that of man. I named this piece of water, *Pelican Lagoon*.[1]

The golden days of the pelican have, for the most part, disappeared, as human populations with their demands for territory have increased, but even during these times there have been some precious and notable exceptions. In 1870 the beauty of the pelican helped in the debate to establish National Parks in the United States. During the discussion, one of the party, Judge Cornelius Hedges, spoke up: 'Are dollars all that matter? Here we have found marvelous scenery where . . . white pelicans, and many other big, beautiful birds nest . . . Could not this place be set apart forever so that all who come after us can enjoy it?'[2]

Thirty years later President Theodore Roosevelt played a role in helping the plight of brown pelicans. German immigrant Paul Kroegel, who lived on the Indian River Lagoon in Florida, was disturbed by the declining population of brown pelicans that lived on a small island near his home. This decimation was due to the exploits of feather hunters (known as 'feather seekers').[3] Feathers were being collected in great numbers for use in the fashion industry as hat plumes, and for quills. Writing quills, in use since the fifth century, relied on a steady supply of feathers, which included those of the pelican. A seventh-century poem mentions the use of the pelican for this purpose:

The shining white pelican which sips
with open throat
The waters of the pool once produced one white.[4]

Pelican Islands, Florida. 'City of Refuge' for pelicans, *c.* 1905.

In 1901 the American Ornithologists' Union and the Florida Audubon Society led a campaign to pass legislation, known as the Audubon Model Law, which outlawed plume hunting in the state. This was an important ruling, but tougher measures were needed to protect the pelicans; Kroegel requested that Roosevelt designate the island as a wildlife refuge. Kroegel was visited by Frank Chapman, an ornithologist and curator at the American Museum of Natural History, who learned that Pelican Island was one of the last rookeries of brown pelicans on the eastern coast of Florida. Roosevelt, acting on advice, including that of Chapman, decided that Kroegel's idea of classifying the island as a wildlife refuge was sound, and on 14 March 1903 Pelican Island Refuge became the United States' first national wildlife refuge, with Kroegel taking on the role as the first national wildlife refuge manager.[5] Not everyone possessed Kroegel's vision; these decisions, unpopular with some, led to the murder of two wardens in the years 1905 and 1908. The 1908 murder sparked the nation's conscience and the Audubon Society intensified its campaign against the wearing of feathers.

Many did not see the need to protect, even in the quest for knowledge about particular species. We may wince when reading this account from ornithologist John Audubon, writing about his exploration of the Florida swamps in 1832 in search of brown pelicans. When found, he spent the day shooting, skinning and then drawing them:

> I waded to the shore under cover of the rushes along it, saw the pelicans fast asleep, examined their countenances

The brown pelican, from Audubon's *Birds of America* (1827–30).

and deportment well and leisurely, and after all levelled, fired my piece, and dropped two of the finest specimens I ever saw. I really believe I would have shot one hundred of these reverend sirs, had not a mistake taken place in the reloading of my gun.[6]

Audubon studied the American white in a similar manner; it is hard to admire his exquisite illustrations without the following description being its violent accompaniment:

The pelicans appeared tame if not almost stupid; and at one place where there were about sixty on an immense log, could we have gone twenty yards nearer, we might have killed eight or ten at a single discharge. But we had already a full cargo and therefore returned to the vessel, on the decks of which the wounded birds were allowed to roam at large. We found these Pelicans hard to kill, and some which were perforated with buckshot did not expire until eight or ten minutes after they were fired at . . . A Pelican had been grazed on the hind part of the head with an ounce ball from a musket, and yet five days afterwards it was apparently convalescent and had become quite gentle. When wounded they swim rather sluggishly and do not attempt to dive or even to bite, like the Brown Pelican, although they are twice as large and proportionally stronger.[7]

We know that Audubon was a product of his time, and we can accept, though reluctantly, his actions and, equally importantly, his thoughts, which were based partly on ignorance. Our unease, though, arises from the knowledge that shooting of pelicans continues, but without the legitimization of the slaughter justified on the grounds of ornithological research.[8]

Others hoped to impress the scientific community. John MacGregor, a Scottish explorer on a tour of the Middle East from 1868 to 1869, documented his struggle with a pelican he had wounded:

> He was far too large and awkward as a cargo to carry two miles in comfort, and cutting off his head would be a troublesome operation. So I resolved to make him carry his own body all the way to the camp by chasing him towards it while he swam . . . His head I brought home, but the great black feet which it was thought would dry into a sort of imperishable leather were soon dissolved into a mass of black meaningless jelly . . . The captured head, which has curly feathers, was shown . . . at the exhibition held in summer by the Palestine Exploration Fund.[9]

Of what use is the pelican? Does the pelican have commercial value? Sadly, it does, and as a consequence several species are under threat. Although pelicans were viewed as sacred in ancient Egypt, associated with the goddess Henet, they were also eaten, hunted in the marshes along the Nile. Organized hunting expeditions used cats to flush the birds from the reeds and utilized

Trapper with pelicans, tomb of Horemhab, Thebes, 1420–1422 BCE.

139

lassoes, weighted ropes, bows, arrows and sticks to bring them down. A strange story recorded by Horapollo in the fourth century CE tells how people set about capturing pelicans:

> Although the pelican is capable of laying its eggs in quite inaccessible places it does nothing of the kind. Instead it digs a hole in the ground and deposits its offspring in it. Seeing what the bird is up to the catchers surround the spot with dried ox dung and set fire to it. When the pelican sees the smoke its great efforts to extinguish the fire by flapping its wings only make the fire worse and it sets fire to itself by mistake.[10]

Several species of pelicans are ground nesters, laying their eggs in flimsy nests on the ground. People still set fire to reeds to flush out pelicans, a practice continued in Western Mongolia by herders and fishermen.

In the pelicans' favour, though, is the fact that their flesh is tough and not pleasant to eat. This has meant that while hunted occasionally for food, including for their eggs and young, pelicans have been saved, for the most part, from large-scale hunting. Writing about the American white, Audubon notes: 'its flesh is rank, fishy, and nauseous, and therefore quite unfit for food, unless in cases of extreme necessity.'[11] Modern-day ornithologist Bryan Nelson echoes these words, describing their flesh as 'tough and fishy'.[12]

Audubon observed the use of brown pelicans as a food source for African Americans:

> The Negroes of the plantations on the eastern coast of the Floridas lie in wait for the Pelicans . . . They skin the birds like so many raccoons, cut off the head, wings and

The American white pelican, from Audubon's *Birds of America* (1827–30).

feet; and should you come this way next year, you will find these remains bleached in the sun . . . At home they perhaps salt, or perhaps smoke them; but in whatever way the pelicans are prepared, they are esteemed good food by the sons of Africa.[13]

Perhaps this document is also social commentary; the harsh conditions of poverty-stricken African Americans fitting the brief of what or whom constitutes 'extreme necessity'. In modern times, pelicans are still shot for food; the flesh of adult great whites in Ethiopia and Egypt is sold in markets.

Pelicans are exploited for other products: their pouches are made into containers, their skin turned into leather, their guano (pelican dung) used as fertilizer and the fat of young pelicans converted into oils for traditional Chinese and Indian medicines. These are not new endeavours; exploitation of the pelican has happened, and in some cases continues, wherever they reside. In 1758 Antoine-Simon Le Page du Pratz wrote: 'The sailors kill them along the sea shores where they are always to be found so as to get this pouch in which they place a cannonball and suspend it so as to shape it into a bag in which they place their tobacco.'[14] Native Americans prepared the pouches until the membrane became as soft as silk. These soft skins were sometimes embroidered by Spanish women before being fashioned into workbags.[15]

Pelican oil or fat is believed to cure a variety of ailments, from rheumatism and sprains, to deafness and boils. The extraction of oil from pelicans is an old practice; written records of it being undertaken on a large scale date back over a century: 'In Cochin China, oil is extracted from the bodies of the many hundred thousand pelicans annually slaughtered for their feathers.'[16]

Guano has been used throughout history; the Chincha people utilized that of the spot-billed, and in Africa, the dung of the

23　Anſeris inſtar auem foelix habet India, Piſces,
　　Funiculo ingluuie aſtrictá, quæ prendere nouit.

　Non glutire, ſed in ripas adferre propinquas,
　Euomere, atej India victum conferre perita.

Carolde Mallery ſculp.　　　　Phil. Galle excud.

pink-backed was used. The Peruvian pelican is a large producer of guano, especially on the Lobos Islands, where it is classed as one of the 'big three' producers.[17] Early in the twentieth century exploitation of guano had reached a high level, with nesting areas of pelicans being destroyed. Measures were taken to farm the islands in a controlled way and to create protected nesting areas on the mainland.

The Dalmatian pelican is hunted for its bill by herders in western Mongolia. The mandible of the upper bill is traditionally used by Mongolian nomads as a horse scraper (as well as for tobacco pouches); they believe that using the upper mandible of pelican beaks to groom their horses makes them stronger and

Karel van Mallery, *Indians Catching Fish with the Help of Pelicans*, 1634, engraving.

faster. Pelican bills have high monetary or trading value; in 2007 one pelican bill traded for ten horses and thirty sheep on the black market in western Mongolia.[18] Although pelican scrapers are a western Mongolian tradition (eastern Mongolians tend to use scrapers made out of soft wood), pelican bills are still able to be obtained at some *naadam* festivals, where they are also on display. Traditionally, pelican horse scrapers were family treasures, handed down for many generations, from father to son, and were not for sale. Now they are expensive items, thought to bring prestige to the owners, via those who, for the most part, just want to make quick and easy money. Illegal shooting continues because the monetary gains make it worth the risk.

Pelicans face a number of threats caused by humans sharing or invading their habitat; problems arise on both the ground and above, in shared air space. One example occurred in Australia; due to drought, a large number of pelicans were found flying in vital airspace at Coolangatta, north of Sydney. The authorities had to hire twenty bird-catchers to try to round them up, to avoid them colliding with planes.[19] Overhead powerlines have been, and still remain, a danger for pelicans in many parts of the world. During a two-year period (2006–8) in Victoria, Australia, at least a hundred pelicans collided with powerlines which were situated across the birds' flight path between wetlands. The investigators wrote about the consequences of impaired flight: 'We were shocked to see bird skeletons and bones strewn under the powerlines and across the paddocks. Normally pelicans are a match for all but the most desperate vixen with kittens to feed, but rendered helpless after their encounter with the hardware of modern information technology they were easy prey.'[20]

The solution, installing visual deterrents on the powerlines to reduce or eliminate bird strikes, appears to have been successful in the case above.[21] Erecting markers on electricity powerlines or, preferably, burying them, has been successful in significantly reducing deaths from collision in a number of regions worldwide.[22]

In some countries pelicans are utilized as decoys, in order to help fishermen. In Pakistan pelicans are tethered, and pelican skins fashioned as camouflage 'beneath which to approach other aquatic birds'.[23] Pelicans, however, are more often the unfortunate victims of fishing line and hook entanglements. In Florida 90 per cent of pelicans receiving rehabilitation are the result of injuries sustained from fishing lines, nets and hooks. Experts believe that 80 per cent of pelicans will fall victim to either an active line or discarded tackle at some point during their lifetime.[24] Bills, legs and wings are hooked or entangled, meaning, for many pelicans, a painful, slow death:

> Little was sadder than the dead pelican I saw hanging from a tree branch after losing the fight to free itself. The line had cut through the flesh on its webbed feet. It had cut through wing tissue, and it had pulled his beak open in a monstrous gape. He had hung there, bound and trapped, for many days before he died, simply because man had been too lazy to discard his line properly.[25]

This common occurrence has led to education programmes in a number of fishing communities, informing fishermen of the risks to seabirds and of the importance of reporting injuries as soon as possible. Sometimes seeing these injuries spurs individuals to action. One such activist was Lance Ferris, known as 'Pelican Man', founder of Australian Seabird Rescue, whose volunteers have rescued over a thousand Australian pelicans. For fifteen years, until his death in 2007, Ferris saved hundreds of pelicans and trained and mentored pelican and seabird rescue teams. His partner, Marny Bonner, says, 'Lance has always had huge compassion for Australia wildlife . . . It wasn't until he noticed not one but two pelicans with hooks in their legs, on the same day, that he realised that there was a problem out there. So he borrowed a boat and had a look around the Richmond River and found that 37 out of 100 pelicans were injured.'[26]

Ferris was overwhelmed by the potential enormity of the problem. He had looked at one island and estuary, and found a high proportion of entangled pelicans; what about the rest of the coast line? He realized early on the need for an extensive education programme.[27]

Until recently, fishing injuries were thought to be associated with discarded fishing tackle and hooks, but results from a 2014 study by researchers from the University of Adelaide's School of Medical Sciences and the Australian Marine Wildlife Research and Rescue Organisation (AMWRRO) suggest otherwise. The study, which involved 113 seabirds treated over a six-year period for 132 fishing-related injuries (pelicans were involved in more than 59 per cent of cases), concluded that, contrary to popular belief, it is not discarded fishing tackle that is the main culprit, but the 'live' tackle used by recreational fishermen; the problem is that pelicans (and other sea birds) are flying too close to those engaged in recreational fishing.[28]

One of the reasons for pelicans becoming entangled or hooked is the decrease in their food source. Climate change, which has led to warmer water, is one of the causes of declining fish supplies, leading to a decrease in the number of pelicans (breeding birds depend on abundant, local stocks of (often) particular species), and an increase in the numbers of starving and emaciated birds. Access to plentiful supplies of anchoveta (*Engraulis*) is a major component for breeding to be a success for pelicans.[29] For example, in Peru, inadequate supplies of anchoveta have played a major role in the decline of pelican numbers.[30] Warmer temperatures mean that anchovetas move into deeper water to stay cool, but pelicans are unable to dive deep enough to reach them. After examining the stomach contents of dead pelicans during a die-off, researchers discovered that some of the digested fish were of species not normally eaten by pelicans, or, even more telling, that the stomachs were, in fact, empty, starvation being the cause of death.[31]

California brown pelicans have experienced unprecedented nesting failures and starved to death by the thousands because Pacific sardines, their most important food, have disappeared. Sardine populations along the West Coast are the lowest since the 1950s due to overfishing and changes in ocean temperature. Starving sea lions, scrounging for offal discarded by fishermen, wound and sometimes kill pelicans, particularly juveniles, as they compete for that waste. Offal, or fish oil contamination, also occurs when pelicans feed at commercial and public fish-processing sites. Pelicans can easily become coated in fish oil as they forage for scraps. Fish oil isn't toxic, but it can be dangerous, because it breaks down feather barbules so they no longer interlock.[32] When an oily pelican is no longer waterproof, it is at risk of hypothermia.

Starvation can lead to unusual pelican behaviour, responses which place the birds at risk. Searching for food has meant that some malnourished pelicans have remained in cooler regions, such as on the Pacific Northwest's Columbia River, for longer periods, rather than flying off to warmer climes.[33] This has meant that weaker pelicans have had to face the ills of approaching winter, the drop in temperature increasing the dangers of frostbite and damage from icy winds.[34]

Starvation has also led pelicans to undertake the unusual practice of killing common murres, or common guillemot. In 2010 pelicans were observed shaking the birds so that they could ingest the regurgitated contents of their stomachs; they would also eat their chicks. In Newport, Oregon, hundreds of murres, killed by pelicans, were found on the shore.[35] During the u.s. sardine famine (2007–14) California brown pelican numbers were extremely low, but there is no record of what those numbers were. According to reports, 'the u.s. Fish and Wildlife Service, in violation of the Endangered Species Act, declined to do proper post-delisting monitoring from 2009 through 2013.'[36] Seabird biologist Laurie Harvey decided to do her own count, with shocking results, because Harvey found that pelican numbers were dangerously low. Before the u.s. sardine famine, numbers were high, with nearly 10,000 chicks recorded in 2006. This healthy record was shattered during the famine, with Harvey recording only five brown pelican chicks on Anacapa: 'In 2012 we had the worst year since 1970 when there were only four chicks . . . But the 1970 failure was directly related to DDT. In 2012 pelicans abandoned their nests; they weren't able to feed their chicks.'[37] It is ironic that the name 'Anacapa' is Chumash for 'House of the Pelicans.'[38] There are hardly enough pelican chicks for a small room, let alone a house.

Sometimes pelicans die because they have consumed incorrect species of fish. In one report, at least twenty brown pelicans died trying to swallow introduced South American sailfin armoured catfish off Puerto Rico.[39] Since 1990 this exotic fish from Venezuela, probably introduced by aquarium hobbyists, has been present in at least eight rivers and two reservoirs.[40]

Pelicans have been blamed for eating fishermen's catch, but in fact they eat very little of what is deemed suitable for commercial fishing. In some cases pelicans are extremely helpful, for they eat introduced European carp that cause damage in waterways. Even though the majority of pelicans do not steal fish that is advantageous to fishermen, many have been persecuted over the years and continue to be hunted. These accusations against the pelican are not new; during the food shortages following the First World War commercial fishermen claimed pelicans were decimating their industry and slaughtered them by the thousands, their nests raided for eggs. In 1918 commercial fishermen claimed that pelicans on and around Pelican Island, the United States' first national wildlife refuge, were eating commercial fish and should not be protected. A federal investigation took place. Alongside scientists and representatives from the Florida Audubon Society, Paul Kroegel also was present, putting forward convincing evidence to demonstrate that the pelicans were not the reason for the decline in fish numbers; rather, the problem was caused by overfishing.[41]

These days, pelicans' opponents include owners of commercial fish farms. Although American whites are protected by the Migratory Bird Treaty Act, shootings in the United States have increased in recent years because the pelicans are seen as a threat by those managing catfish farms. Pelicans are also perceived as a threat by farmers in other countries. In Israel hundreds of pelicans, flying their migratory route across the country, are often shot as they land to feed.[42] 'Israeli wildlife authorities try to avert conflict between the farmers and the birds; measures include establishing alternative ponds for the migrating pelicans' detection, managed by the Israel Nature and Parks Authority.'[43]

Relations, though, are often strained between the Israel Nature and Parks Authority and the owners of commercial fish ponds – so much so that in 2013 two northern coast kibbutzim were suing the Israel Nature and Parks Authority for 5 million shekels (approximately $1.3 million) in damages.[44] They accused the Nature and Parks Authority of neglecting to establish alternative ponds for the pelicans for the 2009–10 season. According to the suit, the Nature Authority decided it would not establish the alternative ponds for the pelicans ahead of time, only doing so if needed. The kibbutzim argued that pelicans had landed in Israel and, finding no alternative ponds, had been feeding on the fish in the commercial ponds for a fortnight.[45]

A dead pelican, having died of starvation, on a beach in Lima, Peru, 2012.

Some of the wounded pelicans end up at the wildlife hospital run by the Ramat Gan Safari Park and the Nature and Parks Authority. One brought in for treatment in December 2013 shocked staff; X-rays revealed that the pelican had no fewer than

110 shotgun pellets in its body.[46] The pelican was suffering from severe lead poisoning, a cracked wing and had lost one eye. The hospital resettles disabled pelicans in designated fish ponds; they are fed for the rest of their lives and their presence attracts migratory pelicans, which authorities hope will land there, rather than at commercial fish farms.[47]

Unfortunately, persecution continues in other places as well, often at the hands of fishermen. An American writer wrote of her surprise at seeing pelicans in St James's Park, London. She had grown up on the Californian coast, where she had seen many brown pelicans: 'I remember hearing as a child about brown

White pelicans, Florida.

pelicans starving to death because someone had sawed off their top bills . . . I recall all the other stories I'd heard growing up of injuries inflicted on the "pesky pelicans" . . . I looked at the beautiful, white birds before me and thought of what a different life they must lead.'[48]

The year 1983 seemed to be a cruel one for pelicans. Late that year there were a number of brutal attacks on pelicans along the Californian coast. In December, 31 pelicans were electrocuted, to keep them off a barge. Even though many fishermen were annoyed at the opportunistic pelicans, this spate of cruelty shocked them. Many suggested that the decline in anchovies was the reason for the pelicans persistently trying to catch fishermen's bait and/or their catches.[49]

Pelican cruelty continues. Nearly ten years after the previous reports of large-scale mutilations, pelicans were again targeted, the attacks stretching along the entire southern coast of California. As in the previous attacks there appears to be a correlation between El Niño waters reaching the California coast, leading to a decreased supply of fish for the pelicans, and a rise in fishermen's anger. Starving pelicans come closer to humans; groups of pelicans congregate on piers, and some even try to eat bait straight off fishing hooks. Their unwanted presence has led to some exceptionally brutal retaliations. One of the worst cases of cruelty towards a pelican was the crucifixion of a live, young pelican on top of a light pole.[50] During these violent attacks, injured pelicans were admitted to various animal shelters, sometimes stretching these centres' resources. For example, the number of injured pelicans cared for by the Pacific Wildlife Center in Laguna rose from three pelicans a week to fifteen pelicans a day.[51]

Torture elicits responses from the outraged. In 1998 the American conductor David Woodard conducted a requiem that included music he had composed for the occasion. The service was held in order to remember and mourn the brutal murder of a California brown pelican on Long Beach. Reports of cruelty continue; in 2015 ten pelicans were killed by having their throats slashed, and a further four injured, in southern Florida, a year after dozens of pelicans were killed in the same brutal manner.[52] Cruelty is not confined to the United States, but is a global problem. On islands in Lake Manitoba and Winnipeg, large numbers of American white pelicans are slaughtered: 'In 1994 more than 2,000, including young, were killed on Winnipegosis and in 1995 more than 600 on Gull Island in Lake Manitoba.'[53] Reported atrocities against pelicans in Canada have included the use of flame throwers to burn live adults and young.[54] In Australia, International Bird Rescue was informed that two Australian pelicans were used as target practice, with one being euthanized as a result of its injuries.[55]

In recent years research has been undertaken in the field of prosthetics, constructing artificial bills for pelicans, in order to heal injuries sustained from violent human acts. Most, if not all, of these attempts have been unsuccessful; early attempts to replace the bills surgically with fibreglass replicas failed when the bones to which they were attached deteriorated.[56] Dr Gayle Roberts, one of the vets involved in the first surgical attempt, said the pelicans would starve to death without the upper beak, which is needed to hold the fish they scoop up with the bottom beak. Pelicans suffer extreme pain when their beaks are chopped off but they continue to try to catch fish, unsuccessfully.[57] Recently there has been a glimmer of hope; in 2015, 3D printing technology entered the picture, creating an artificial bill for a white pelican with a broken beak at the Dalian Forest Zoo, northeast China. The zoo staff, after hearing about 3D printing being used at a city hospital to print out patients' bones for surgical reference, decided to see if this technology could create an artificial bill.[58] The beak seemed to show some indications of new growth, so vets preserved that portion of the bill and added a 3D printed partial beak extension, which allowed the bill to be full-size. This 5-mm-thick bill appears to be a success; on 17 August the pelican had begun feeding on his own for the first time since his injury back in May.[59]

Another pelican who had reconstructive surgery for a broken beak received more coverage than was warranted. Pierre the Pelican, the New Orleans NBA mascot, needed his beak repaired after a major injury. Press releases included an amusing statement from Carolyn Atherton, the curator of birds for the Audubon Zoo: 'We do see a lot of head trauma cases with pelicans,' Atherton said. 'We've had quite a few pelicans we've treated for injuries similar to Pierre's. After they've been treated, they've all come back looking bigger, stronger, faster and ready to take on the world.'[60]

In April 2013 the New Orleans Hornets officially changed their moniker to The Pelicans. The mascot's new beak may have been an attempt to address criticism of the original design, described as 'creepy'. The original design had become the subject of some mock-up horror movie references; following the unveiling of the mascot, Pierre the Pelican, Twitter users began posting jokes about the mascot's alarming appearance, including digitally editing images of the mascot superimposed onto scenes from a number of horror movies.[61]

Pelicans are commonly the victims of human error (oil spills, poison and chemical build-up), or of human intent (hunted or maimed). Poisons, in the form of pesticides, heralded both the near extinction of the brown pelican, and a late twentieth-century success story for the conservation movement. Between the late 1950s and the early 1970s brown pelicans were on the edge of extinction in North America, disappearing entirely from the Pelican State of Louisiana, due to pesticides entering the food chain. In 1970 the FWS (U.S. Fisheries and Wildlife Services) listed the brown pelican as endangered. The pesticide endrin killed the pelican outright, while the chemical DDT (dichlorodiphenyltrichloroethane) produced thin-shelled eggs. Residues of DDT in the fish the pelicans consumed were believed to have prevented the mothers from depositing calcium in the shells of their eggs, which caused eggs to break easily under the weight of the parents, thereby producing no live young. A 1972 ban on DDT, along with transplanting thousands of chicks from Florida to Louisiana, led to remarkable population recoveries. In 1983 several pairs of brown pelicans nested on Gaillard Island, the first-known nesting site in Alabama. In 1985 Atlantic coast populations of the brown pelican were removed from the endangered species list. By the late 1990s brown pelicans reached pre-pesticide numbers; successful breeding led to abundant numbers in 2009. That year heralded exciting news concerning the surge in pelican numbers, which included 70,000 breeding pelican pairs in California, and approximately '620,000 pelicans inhabiting the West Coast, Gulf Coast and Latin and South America.'[62] The species was fully delisted in November 2009, ironically, just five months before the Deepwater Horizon oil spill. An updated version of Merritt's classic celebrated this victory.[63]

The American white pelican also experienced thinning shells during the 1960s, but by 1997 the thickness of the American white's egg was nearly back to its pre-1947 measurement.[64] Pesticides were used in other countries, too, leaving harmful concentrations of DDT, DDE (dichlorodiphenyldichloroethane) and hexachlorocyclohexane at high levels in great white pelican eggs and chicks in the Ili Delta, southern Kakahstan, during 1988 to 1989.[65]

The increased interest in ecology has made a difference to the survival of pelicans, but the effects of chemical poisoning continue, with different chemical threats replacing DDT. Of all the known biotoxins, the least understood is probably the harmful algal blooms (HABS). Nitrates from fertilizers and raw sewerage are washed into seas and rivers; these feed algal blooms that leach oxygen from the water. The HABS produce a potent neurotoxin, domoic acid (DA), from diatoms. This DA is transmitted to marine birds and mammals via fish such as northern anchovies (*Engraulis mordax*).[66] Fish become ill with botulism, and within hours of eating infected fish, pelicans can suffer from a range of clinical signs that include seizures, paralysis (which can be so severe that they are unable to hold their heads up, and thereby drown), blindness and death.[67] Outbreaks of avian botulism have occurred in North America since the early 1990s and have caused the deaths of millions of birds.[68] One example to illustrate the magnitude of the problem, happened near Baja California in 1996. Large numbers of brown pelicans died, the cause of their death attributed entirely to the ingestion of fish contaminated by DA-producing diatoms.[69] Recently in Australia, the New South Wales Environmental Protection Authority launched an investigation into the unexplained loss of hundreds of pelicans on the Central Coast since 2014, concluding that the deaths were linked to harmful algal blooms.[70]

During the winters of 2008–9 and 2009–10, brown pelicans in California, Oregon and Washington State displayed symptoms of disorientation, fatigue and bruising. Many of the brown pelicans were found on highways, airport runways and on farms, often some distance from the coast. At the beginning, several were tested for signs of DA poisoning, but this toxin is usually a problem in spring and summer, not in winter; other toxins did not appear to be the culprits either. Post-mortem examinations revealed that starvation, malnutrition and anaemia were major factors that contributed to the deaths of the examined pelicans;[71] some of the items in their digestive tracts were of fish atypical of their diet, such as sea urchins, squid and marine worms, which indicated a shortage of northern anchovy and Pacific sardine. Severe weather conditions also played a role in 2009; half of the stranded pelicans who died had frostbite injuries.[72] Malnutrition and weather-related stresses could contribute to increased susceptibility to disease.

Even though things were tough, brown pelicans were slowly making their way back to pre-pesticide numbers and were being fully delisted. Unfortunately, this was the time when disaster struck in the form of the Deepwater Horizon oil spill, which threatened the brown pelican Gulf Coast populations anew, and brown pelican status again. Since brown pelicans breed, roost and forage near shipping channels, they are highly susceptible to oil spills. In the summer of 2010, brown pelicans, covered in oil, became the iconic image, the poster bird, of the disaster.

The Deepwater Horizon oil spill was caused by an explosion on 20 April on the Deepwater Horizon oil rig in the Gulf of Mexico, 66 km (41 mi.) off the coast of Louisiana. The rig was leased by oil company bp (it is ironic that 'bp' could also be an abbreviation for 'brown pelican'). While this was the worst offshore oil spill in u.s. history (and therefore the worst marine environmental disaster in u.s. waters), it was not the first oil spill, and, sadly, will not be the last. Since 1964 over 320 known spills have occurred in the Gulf due to offshore drilling.[73]

By the time the well was capped on 15 July 2010 an estimated 134 million gallons of crude oil and 4 million pounds of gas had escaped into Gulf waters.[74] Oil was not the only problem; the 2 million gallons of toxic dispersants used to break down the oil may have contributed to the toxicity levels in the ocean, making the water more dangerous for marine life. The action of the chemicals in the dispersants does not remove oil, rather it enables the oil particles to break down into smaller components, to 'disperse', rather than to 'disappear'. Ironically, this may increase the toxicity of the oil because its smaller size means easier access to the food chain.[75] The petroleum formed a slick; an estimated 1,770 km (1,100 mi.) of shoreline was polluted, including coastlines and beaches of Louisiana, Texas, Mississippi, Alabama and Florida.[76]

Birds were particularly vulnerable to the effects of the oil, and many perished, either from ingesting oil as they tried to clean themselves or because the substance interfered with their ability to regulate their body temperature.[77] The beautiful iconic brown pelican of the coast, flying above the waters, overnight became the archetypal oil-soaked pelican, grounded, unable to fly.

Several months after the disaster, brown pelicans were acknow-ledged as the bird species that suffered the most from the oil spill.

Pelican being cleaned as a result of the California oil spill, 2015.

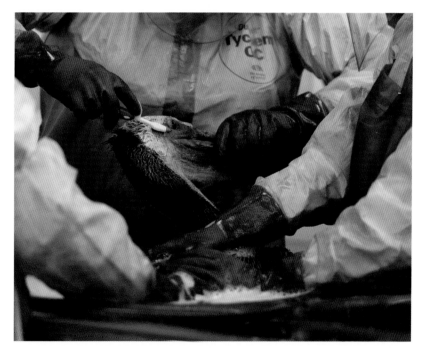

Jeff Phillips, from the u.s. Fish and Wildlife Service, rescues a brown pelican from the Barataria Bay in Grand Isle, Louisiana, 4 June 2010.

u.s. Fish and Wildlife Service statistics report that '58 per cent of all dead or injured birds collected up to that point were pelicans.'[78] Brown pelicans were also casualties of bad timing, because the spill happened at the beginning of their annual nesting season. Biologist Kelly Hornaday of the u.s. Fish and Wildlife Service estimates that 'between 25,000 and 33,000 pairs of pelicans breed on islands off the u.s. Gulf Coast; Louisiana alone hosts between 8,000 and 16,000 pairs, depending on the year.'[79]

Approximately 6,000 birds were found dead the first year after the Deepwater Horizon oil spill. The official number represents only the number of birds collected by wildlife officials. Scientific research indicates that mortality can be assumed to be four to eleven times higher than the number of birds retrieved, a common estimate being ten times higher.[80] In the year after the spill '932 brown pelicans had been collected, so we could assume that more than 9,300 had been harmed'.[81] According to a 2014 study, up to 800,000 coastal birds were thought to have died as a result of the oil spill during its acute phase.[82] A 2015 report stated this loss included approximately 12 per cent of the population of brown pelicans in the northern Gulf of Mexico.[83]

Despite good intentions, cleaning oiled pelicans does not necessarily save their lives, and cleaned pelicans that do survive may never be able to reproduce: 'Birds such as the brown pelican can ingest oil while cleaning their feathers or by consuming contaminated food. Oil exposure can lead to long-term physiological, metabolic, developmental or behavioral effects, which can in turn lead to reduced survival or reproduction.'[84] Clean-up efforts only remove a fraction of the persistent oil and gas spilled; millions of gallons remain in the Gulf.[85] In 2012 researchers found traces of pollutants (petroleum compounds and remnants of a dispersant compound, the chemical Corexit, used to break up oil spills) in the eggs of American white pelicans in three Midwestern states: Minnesota, Iowa and Illinois.[86] The results are troubling because these contaminants could cause birth defects in developing pelican embryos. The Environmental Protection Agency states that Corexit contains carcinogens as well as endocrine-disruptors, both of which could interfere with growth hormones during an embryo's development. 'Nearly 80 per cent of the eggs tested so far contain Corexit, while 90 per cent contain petroleum compounds.'[87]

Another major concern for the recovery of brown pelican populations is the ongoing erosion of coastal barrier islands and marsh habitat. One example is Cat Island, a barrier island and nesting spot for brown pelicans in Barataria Bay near Grand Isle, Louisiana, which was protected with inflatable booms after the oil spill in 2010. The island is now almost entirely under water and some scientists suspect that oil killed its mangroves, speeding erosion.[88] No one knows where the Cat Island pelicans have gone but the disappearance of the island would pose problems for brown pelicans, who usually return to their natal site.

Grief may find expression through creative outlets, or rituals, as already noted in the example of David Woodard, who composed music for a memorial service in 1998. After the Deepwater Horizon disaster, plans were laid for a way to remember the destruction of wildlife. On 5 June 2010 in New Orleans, Ro Mayer, a real estate agent and costume designer, led a jazz funeral parade

Brown pelicans covered in oil after the BP oil spill, at a rehabilitation centre waiting to be cleaned, June 2010.

'Dead Pelican Sandwiches', satirical second line parade in response to the BP Deepwater Horizon oil spill disaster, 5 June 2010.

in traditional New Orleans style to commemorate the dead. Mayer, the drum major for a group she called the 'Krewe of Dead Pelicans', carried a pelican-adorned umbrella.[89]

In other regions of the world, pollution can lead to visible, physical changes in pelicans. In May 2016 several black pelicans were photographed on Penguin Island, Australia; this spotting created global excitement in the world of twitchers and wildlife enthusiasts. The three dark pelicans could fly, showed no visible signs of distress and seemed to be accepted by other pelicans in the colony. Because more than one pelican was affected, this pointed to pollution being the cause, rather than an overproduction of melanin.[90]

What of the other species? How are they faring? Five of the eight species (great white, Dalmatian, American white, brown and spot-billed) have declined in the past century. The most endangered species are the Dalmatian and the spot-billed. The Dalmatian, declining because of hunting, degradation of wetlands and colony destruction by fishermen, once numbered in the millions. The population is now estimated by the IUCN to consist of between 10,000 and 13,900 birds globally: 4,300–800 in the Black Sea and the Mediterranean, 6,000 in southeast Asia and south Asia, and fifty in east Asia.[91] The Dalmatian pelican is classified as Vulnerable on the IUCN Red List 2016, with the easternmost population, in Mongolia, at a higher level of risk.

Khar-Us Lake National Park is one of the last breeding sites for the Dalmatian pelican in Mongolia. Some of their nesting places have disappeared, due to physical damage caused by the muskrat *Ondata zibethica*, introduced to Khar-Us Lake in 1967 for fur production.[92] 'Since then, the muskrat has spread widely, invading lakes and the wetland ecosystem.' Livestock grazing has also had a negative effect, the reed beds being used as grazing areas for large numbers of livestock, especially during the winter. In January 2006 researchers noted 113 herder families and over 22,000 livestock using the area.[93] Uncontrolled fishing and over-fishing in the lakes is also a problem, plus the increasing practice of setting fire to the reeds by herders and fishermen.[94]

Unfortunately, the Dalmatian pelican has also been a casualty of avian flu. In March 2015 bird flu virus H5N1 was detected in 64 carcasses of pelicans in Romania's Danube Delta. Another

report put the count higher, at over one hundred dead pelicans.[95] Regardless of which is the correct number, the damage was extensive, because that colony was small, consisting of approximately 250 birds.[96]

Yet there are hopeful signs. Considered extinct since the 1950s in Croatia, a single Dalmatian pelican was spotted on the mouth of the Neretva River in 2011.[97] Dalmatian pelicans had been absent for decades, due to uncontrolled hunting and fishing activities in the region. During severe climate conditions, humans have sometimes stepped in to help. The winter of 2012 was unusually cold in Europe, with the Caspian Sea freezing over. In the southern Russian province of Dagestan, authorities stepped in to save hundreds of rare Dalmatian pelicans, trapped by the freezing conditions. Dagestan's Nature Protection Ministry had hundreds of kilograms of fish brought in every day for the hungry pelicans, and volunteers hand-fed the birds, enabling them to survive the harsh winter.[98]

For all pelican species, a pressing issue is loss of habitat. Preservation of adequate breeding and feeding habitat remains paramount. Due to increased extraction of water, some lakes and rivers have dried up completely in recent years, including Lake Oubeira in Algeria, Lake Marmara in Turkey, and the Axios, Nestos and Strymonas rivers in Greece. Rampant coastal development

Dalmatian pelicans in Divjaka-Karavasta Lagoon National Park, Albania.

has destroyed traditional pelican nesting sites, some of which contained thousands of nests. Conservation efforts have been undertaken, especially in Europe, to address some of these dangers. Although Dalmatian pelicans normally nest on the ground, they have been encouraged to nest on platforms erected in Turkey, Greece, Bulgaria, Romania and Mongolia.

The spot-billed pelican, once numbering in the millions in Myanmar, is mainly restricted to Sri Lanka and southeast India; it is classed as Near Threatened, for the species is declining at a moderate to rapid rate, with the population estimated in 2016 to be between 7,000 and 10,000 in South Asia, and 13,000 to 18,000 globally.[99] A crucial factor in its decline was the loss of the Sittang valley breeding colony in Myanmar through deforestation.

The European-Asian population of the great white is currently threatened, but not the African population at this stage. The European population is estimated at 4,900 to 5,600 pairs, using ten to thirteen breeding grounds.[100] To lose even one significant breeding ground or region can have dire consequences. In the case of the great white, most of its prime area, the Turkish wetlands (Sultansazligo), has been destroyed, primarily by drainage (a human cause). This wetland destruction decreased the great white population greatly, estimates ranging between 50 and 98 per cent.[101] Many other vital sites are threatened.

According to the 2015 Waterbird Conservation for the Americas, the global breeding population of the American white is in excess of 120,000 birds. On the IUCN scale this classes them as of Least Concern, but on the Continental Concern Score they rate as of Moderate Concern. Almost half the American white pelican population breeds at four sites in the northern plains. In 2002 breeding success was low; chick mortality, which had been less than 4 per cent, was at the staggering rate of 44 per cent.[102] The reason for this was illness, in the form of the West Nile virus (WNV).

The pink-backed and Australian pelicans have large numbers and stable populations, so are viewed as of Least Concern. The brown pelican too is listed as being of Least Concern, but this rating is being questioned. The Peruvian was listed as Near Threatened by the IUCN in 2016, due to the decline in anchovy numbers.[103] Illness has also been a factor; in April 2012 the government of Peru began investigating the deaths of more than five hundred pelicans along a 70-km (43-mi.) stretch of its northern coast; their deaths were probably caused by a virus.[104]

Although pelicans are not domesticated birds, some have been partially tamed, and have lived alongside humans. One of the most famous was Parsifal, also known as Monsieur le Pélican, who was rescued by Dr Albert Schweitzer. Schweitzer went on to write *The Story of My Pelican*, a slim volume, very much a photographic study, with commentary provided by Monsieur le Pélican. Parsifal was one of three orphaned fledgling pelicans brought in for money by the hunter who had killed their mother. In real life Schweitzer faced the kidnapper: 'Don't you know it's a sin to take little ones away from their parents? Wait and see, the good Lord will find a way to punish you.'[105] Schweitzer nursed the pelicans back to health. Parsifal, however, slower in his development than the other two, stayed behind when the others finally left.

When Charles R. Joy, a Unitarian minister and international humanitarian worker, visited Schweitzer, he was introduced to Monsieur le Pélican:

> There, above our heads, perched on a high trellis gate, was a huge pelican.
>
> 'Bon soir, Monsieur le Pélican!' said the doctor.
>
> And the pelican flapped his great wings, thrust his long beak up in the air, and with several harsh, throaty expulsions of breath greeted his friend, the doctor.

Dr Albert Schweitzer and his hospital in Lambaréné, Gabon, in 1953, with pelican.

'He is my night watchman,' said Schweitzer . . .'While he is there during the night, no one can pass up these steps except Mlle. Emma . . . and myself. Everyone else gets a powerful rap on the head.'[106]

Whenever Schweitzer returned to Europe, he always took with him a miniature watercolour of Monsieur le Pélican, which he placed on his desk.[107]

Producer and director Judy Irving's film *Pelican Dreams* (2014) examines what it means to be wild. Irving has been fascinated by pelicans for many years; the name of her production company, Pelican Media, reflects this passion. Irving had ceased her research on pelicans in order to make the 2003 documentary *The Wild Parrots of Telegraph Hill*, but returned to them in *Pelican Dreams*.

Pelican Dreams is a documentary about brown pelicans, its primary focus the lives of two pelicans undergoing rehabilitation. One, an emaciated, four-month-old brown pelican that Irving named Gigi, was found in 2008 on the Golden Gate Bridge. This pelican, which became famous on YouTube, caused a massive traffic jam until he was rescued by a truck driver and taken to International Bird Rescue for rehabilitation. The other, named Morro, is cared for by rehabilitators near Morro Bay, central California. Morro's permanent wing injury means he is yard-bound and unable to join pelicans in the wild.

The documentary explores pelicans' nesting grounds and survival challenges, including climate change, the fishing industry and pollution. Ultimately, though, the film is about wildness. Irving, through the subject of the pelican, asks questions about wildness, and wild animals: 'How close can we get to a wild animal without taming or harming it?' Irving then turns the tables on the viewer, concluding with a question for our own (human) souls: 'Why do we need wildness in our lives and how can we protect it?'[108]

Irving dares to ask about the pelicans' feelings: 'Are they lonely?' 'What are they teaching us?' 'Can they dream?' Irving's desire for an emotional connection is in sharp contrast to the scientific approach of Gigi's rehabilitator, Monte Merrick. When Irving suggests that he refers to Gigi by name, not by the band number, he declines, stating that the reason for this is 'so that we remember that they're not pets'. Even when pelicans have recovered at the rehabilitation centre and are set for release back into the wild, Merrick does not join the farewell journey, preferring to stay back at the rehabilitation centre, perhaps keeping emotional as well as geographical distance.[109] Do we allow wildness, or do we live in the hope of taming, subduing the undomesticated into a product that we humans can own?

Flinders was, indeed, prophetic. The golden age of the pelican has disappeared, idyllic days darkened by the storms of hunting, pollution, cruelty, starvation and habitat destruction. Perhaps we humans can recover and repair some of the remnants of glories past, so that pelicans can live their wild lives in safety. Dare we

Great white / Eastern white pelican (*P. onocrotalus*) standing on a bench in the morning sunshine with people in the background, St James's Park, London.

listen to the voice of the pelican? 'Monsieur le Pélican is, like the Doctor, a preacher. On the veranda of the Doctor's house he takes his place, and makes an oration, gesturing vigorously with his wings, and making harsh, throaty noises as he opens his bill wide. He, too, has a mind of his own, and a message to the world.'[110]

Conclusion

The pelican narrative is a rich and varied one. Some soar and glide, but others plunge dive, burdened by the weight of oil, cruelty and the hunt. Still, throughout all its seasons, from its golden age until the present, the pelican continues to intersect with the human. Its presence in wilderness, culture and the human psyche continues to elicit expression through art, literature, engineering, the garish or kitsch . . . a symbol in Christianity, a catalyst for grief support and detected within the folds of paper plane design!

Irony, though, is one of the understated ingredients in this mix. From being classified as 'unclean' in some versions of the Hebrew Scriptures (the Old Testament), to its spiritual elevation in Christianity, symbolizing Christ, is quite a turnabout. But irony is not confined to its association with religion, it makes its presence felt deeply in the secular. Louisiana, known as 'the pelican state' was, for a time, 'pelican-less'. Although the dangers of DDT (dichloro-diphenyltrichloroethane) have been addressed, other poisons and pollutants affect pelican numbers, be it through disease, oil or starvation.

Ironic, too, that Pelican Island, the site that sparked the beginnings of the conservation movement in the United States, created to be the first National Park in order to protect its pelicans, has been at risk. This time the threat was not from plumage hunters,

Pelican's neck in artistic bend.

Within the image:
ELEVEN FIRST PRIZE MEDALS.

GEO BARKER.

NIAGARA FALLS, N.Y.

AMERICAN FOREIGN · GENRE & COMIC GEMS OF INSTANTANEOUS PHOTOGRAPHY

GRAND PRIZE DIAMOND BADGE 1867

GOLD MEDAL PARIS 1869

Florida—Pelican Island, Indian River.

Copyright 1891 by Geo. Barker.

Pelican Island, Indian River, Florida, *c.* 1891.

but from commercial interests. The 1960s heralded the much greater risk of and damage from developers. Their plans threatened the protection, and therefore the numbers, of pelicans. The wetlands and islands had been coveted, viewed as prime real estate, an ideal location for future housing development. Following in the footsteps of Paul Kroegel, locals took action. In 1963 Pelican Island was designated a National Historic Landmark, and in 1970 it became the smallest wilderness area in the National Preservation System. These decisions sound reassuring, but this iconic landmark, the sparkling jewel of the National Parks genesis, is shrinking. By 2000, owing to erosion, it was less than half its original size. Unfortunately, it is not only the land mass that is diminishing. In 1903, when the Refuge was established, over 10,000 brown pelicans were counted at the peak of the breeding season; nowadays the numbers have severely dwindled, declining to approximately a hundred nesting pairs during the spring. Although conservation measures are in place, including the planting of mangroves and hardy cordgrass, we hold our breath and cross our fingers.

Pelicans are wild creatures, a fact which usually limits frequent, ongoing interaction with humans. Sometimes, though, circumstances allow humans not only to acknowledge or study pelican magnificence, but to assist those in need. In February 2014 a young great white pelican, separated from its flock during a storm, was rescued by the camp manager at Greystoke Manhale safari park in Tanzania. The young pelican lacked two crucial skills necessary for survival: knowledge of fishing techniques and an ability to fly. The staff devised ways to teach the pelican both these skills. A camera on his beak recorded the lessons.[1] Viewers witnessed improvements in the pelican's skills, and perhaps experienced the joy which can emanate when observing another species. The team was successful, though it took some effort to teach him how to fish. Even though the pelican, named 'Big Bird', can now fly and fish, and is free to leave the area, he is from a communal species

An Australian pelican (*P. conspicillatus*) in flight.

175

Pelican with folded wings.

and has chosen to stay at the safari park with his adopted 'flock'. The 'wild' has been welcomed and saved. Perhaps these humans were imitating the charitable quality found within variants of the legend of the pelican in its piety: caring for its young.

Although pelicans have often been, and still are, portrayed as slightly comical characters, due to their squat build, large beak and gular pouch, there is nothing amusing about the dangers they face. Pelicans are still at risk, some species more than others. We humans need to clean up our planet so these majestic birds, with a lineage dating back at least 40 million years, may continue to grace land and sea, our minds, our deep dreams and our shared airspace: 'And to lose the chance to see . . . a file of pelicans winging their way homeward across the crimson afterglow of the sunset – why, the loss is like the loss of a gallery of the masterpieces of the artists of old time.'[2]

Timeline of the Pelican

146–100 million BP

Two pelican-like pterosaur fossils uncovered in Chaoyang Liaoning province, China, from the Early Cretaceous era. Named *Ikrandraco avatar*

Eocene epoch (40 BP)

Fossils similar to pelicans found in Eocene deposits

30 million BP

Fossilized beak found in France; lack of evolutionary change for the past 30 million years

1664

Pelicans are introduced to St James's Park, London, a gift from the Russian Ambassador to King Charles II

1860

The pelican is featured on the earliest state flag of Louisiana

14 March 1903

Florida: Pelican Island Refuge is created by President Roosevelt

1937

Pelican becomes the new non-fiction imprint of Penguin Books.

1993

The world's largest paper plane, built by NASA and high school students, is named the White Pelican

1996

In California massive mortality occurs in brown pelicans after they feed on mackerel contaminated by domoic acid-producing diatoms

2007

Peruvian pelican, *Pelecanus thagus*, is defined as a separate species, and not as a subspecies of the *Pelecanus occidentalis*

November 2009

The brown pelican is removed from the Endangered Species List

Miocene epoch (20 million BP)	2200 BCE	1550 BCE	Mid second-century CE
Pelican fossil, *Pelecanus gracilis*, from the early Miocene, unearthed in France	Excavations in the mid-1990s at a site in Peru, Caral, unearthed 32 flutes made of pelican wing-bones	In ancient Egypt, the pelican was known as or affiliated with the goddess Henet. Pelicans are mentioned in the *Book of the Dead*	The book *Physiologus*, written in Alexandria, Egypt, narrates the legend of the pelican feeding her young with her blood

1957	1963	1970	1984	1985
Pelican Pete, at Pelican Rapids, Minnesota, is the world's largest pelican statue	*Storm Boy* by Colin Thiele is published	The brown pelican is placed on the U.S. Endangered Species List, remaining there until 1985	New Orleans hosts the Louisiana World Exposition. The official mascot is Seymore D. Fair, a pelican	Japan's first artificially incubated pelican, a great white, was born in Tokiwa Park, Ube City

20 April 2010	2013	March 2015	12 June 2015
Deepwater Horizon oil spill in the Gulf of Mexico	Three new great white pelicans join the descendants of the original Russian flock in St James's Park, London	H5N1, bird flu virus, is detected in two dead pelicans in Bulgaria. H5N1 is also detected in dead pelicans in Romania. Reports range from 64 to over 100 casualties.	Brown pelicans are fitted with GPS devices following the Refugio oil spill in Santa Barbara on 19 May 2015

Appendix

GROUND NESTERS

Great White Pelican: *Pelecanus onocrotalus*

The great white, at 15 kg (33 lb), is the heaviest of all seabirds (the Dalmatian is the largest).[1] It has big black wing feathers, visible from beneath when in flight, and a white tail. The adult male, pre-breeding, is almost entirely creamy white, often with rosy tinges, particularly on its wing-coverts. Some have yellowish patches on the breast. A ragged crest (10–14 cm (4–5.5 in.)) of long feathers graces the nape (adult non-breeders lack crests), and its white forehead feathers end in a point. The pouch and facial skin are bright yellow or pinkish. Its long, yellowish bill has a central red stripe and a small red nail at its end. The male has a longish beak that grows in a downwards arc, whereas the female's beak is shorter and straighter. There are two populations of the great white: African and Palearctic.

Failure of breeding may be due to desertion (abandoning their nests in self-preservation, or because of adverse feeding conditions), flooding (their nests are on the ground) or predation. Predators include Nile crocodiles in Africa, eagles (which eat eggs, nestlings or fledglings) and sometimes jackals and lions.

American White Pelican: *Pelecanus erythrorhynchos*

The American white pelican is one of the largest North American birds. It has broad white and black wings. Pre-breeding, it may have a slight yellow tinge on its upper wings and breast. Narrow plumes grace its neck and nape, and it sports a pale yellow crest. Its legs and feet are orange, its facial skin is yellow and its pouch is predominantly orange; its bill is also orange or pink. One of its most striking features is a 5-cm (2-in.) fibrous horn on its upper bill: during the breeding season, the American white pelican is sexually ready to mate when this bump grows on the upper mandible (this feature is not present in the other seven species).[2] After breeding, the horn disappears, as do the plumes on the nape, and the bill, pouch and facial skin become pale yellow. Their main predators are foxes, coyotes, gulls, ravens, great horned owls and bald eagles.

Dalmatian Pelican: *Pelecanus crispus*

The Dalmatian pelican is the largest of the pelican species. Its length is between 160 and 183 cm (63 in. and 72 in.), it can weigh between 7.25 and 15 kg (16 lb and 33 lb) (making it one of the world's heaviest flying bird species) and has a wingspan of 290 to 345 cm (114 to 136 in.). Its bill, 36–45 cm (14–18 in.) in length, is the second largest of any bird – the largest being that of the Australian pelican.

It is stunning, with silvery grey feathers, grey legs and feet, an orange-red pouch and a pale yellow bill. The bare skin of the face is yellow, turning purple during breeding. During pre-breeding, it has a top-knot of curly feathers on the back of its head and neck, which led Russians to refer to it as the 'curly pelican'. These curly nape feathers are longer, and quite untidy during the breeding season, giving it a 'bad hair day' look. The loose feathers around the forehead can form a 'W' shape on the face, right above the bill. Post-breeding the Dalmatian pelican's pouch becomes pale yellow, its bill dark grey, and it loses its crest. In flight it holds its neck back, almost heron-like.

Dalmatian pelican habitat is often on islands, where low ground can be flooded in storms. A decrease in water levels due to drought or irrigation can enable predators to cross into these colonies. Predators include wild boars, jackals, foxes, wolves, dogs and lynxes.[3] In some localities the threats are human, with breeding colonies repeatedly destroyed by fishermen or poachers.[4] In addition, colonies are abandoned because of more 'positive' intrusion: tourists, photographers and scientists.

Australian Pelican: *Pelecanus conspicillatus*

The pre-breeding adult male is predominantly white, with a black rump with a white 'V', black upper wings and primaries, and a black tail. His crest is grey, black or white. Breeders develop long pale yellow tufts of feathers at the nape. The female has similar markings. During courtship the bill is pink, with a bluish tinge and a dark blue stripe, the pouch mainly scarlet, rimmed with dark blue. Nuptial colours fade soon after nest-site selection has taken place.

The Australian pelican is allopatric and is widely distributed over most of mainland Australia, and Tasmania.[5]

The Australian pelican is broader in its taste than other species, and will feed on crustaceans such as yabbies and shrimp as well as tadpoles and turtles. During periods of starvation, they have been seen eating other birds, mainly gulls (holding them underwater and drowning them before eating) and ducklings.

The Australian pelican is fortunate, for it has few predators. Predators include Australian ravens and gulls (which attack pelican chicks), dogs and cattle (which can crush eggs). Unfortunately, courting adults can be added to the list as they can accidentally crush eggs and destroy nests during their dance rituals.

ARBOREAL (PARTIALLY/WHOLLY)

Pink-backed Pelican: *Pelecanus rufescens*

The feathers are white, pale grey and pink with dark grey wing tips and a black rump. Its breast and abdomen are pink, and there is a pinkish hue on its back (hence its name). The facial skin develops a black patch, about ten times the size of the eye, with a smaller pink-patch above it.[6] The bill is pinkish, its pouch pink to deep yellow with vertical dark stripes. The pink-backed sports a small crest, which gives the head a peaked look, and the breeding adult has long ragged feather plumes. Its colours are brighter during the breeding season, the pale pink facial skin changing to yellow or orange and its legs becoming pinkish-red. After breeding, the colours fade, the crest shortens, the plumage overall is greyer and the black eye-patch becomes less obvious.[7]

They are gregarious, and will often nest alongside other birds, including the great white pelican. They depend on small fish, so they forage in shallow water; this means that they are not competing against the great white for food (the great white feeds in deeper water further from the shore, catching larger fish).[8] This means they are not viewed as depleting fish stocks by most fishermen, which is in their favour.

Their trees are sometimes close to humans. There have been reports of nests on walls of old African cities such as Kano in Nigeria.[9] Pink-backed pelicans have been observed nesting on the walls of old African cities, including those in Kano.[10] The main threats for the Pink-backed pelican are tree loss and pollution from pesticides.

Spot-billed Pelican: *Pelecanus philippensis*

This small pelican (between 4.1 and 5.7 kg (9 and 12.5 lb)) has a spotted bill, a spotted pink-purplish pouch, a tufted greyish nape crest, and a speckled head and neck. Its wings are dusky grey with dark brown to black tips and dull white to pink undersides. After breeding, both sexes lose some of their facial colour and are browner on the head and back. In the young, the bill spots begin to develop at six months but are indistinct until about a year old.

The spot-billed pelican is a social species, living and travelling in flocks. Breeding populations are confined to India, Sri Lanka and Cambodia. Breeding success is high in this species, with any failure due mainly to human disturbance. In areas where colonies are protected from humans, breeding failure is usually due to two reasons: either juveniles falling out of trees, or suffering caused by extremes in weather conditions, such as high temperatures and cyclones.[11]

Mature spot-billed pelicans have no predators, but the nestlings, fledglings and eggs do, including crows, brahminy kites and jackals. Adults are fiercely territorial, defending their nests from others. The weight and size of brown pelicans varies considerably across species and sub-species. The smallest is the *occidentalis*, and the largest is the *thagus*, or Peruvian pelican. The differences may be due in part to environmental conditions, such as water temperature, which affects fish numbers.[12]

Brown Pelican: *Pelecanus occidentalis*

Brown pelicans have sinuous necks and long dark bodies. Their wings are long and broad, and often bowed when the pelicans are gliding. The adult has chestnut brown feathers, black primaries with white on the shafts, a white neck, some yellow tinges on the head and a pale yellow crown. Nape feathers form a chestnut mane. The bill and pouch are greyish. The lower mandible may be blue, spotted with red, and the bill-tip orange-yellow. There are regional differences: on the Pacific Coast, brown pelicans have red skin on their throats during the breeding season. On the Atlantic and Gulf Coasts, brown pelicans are slightly smaller and their throat skin is greenish black.

The brown pelican covers a wide range, for it inhabits the Atlantic, Pacific and Gulf Coasts of North and South America. The brown pelican is rarely seen inland, for it favours marine habitats, but exceptions are at the Salton Sea in California (where it is fairly common), lakes in Florida and bodies of water in southeast Arizona.

Brown pelicans are social and gregarious; being a colonial species they live in large flocks during the year. They usually feed close to the colony, singly or in groups, but, unlike other species, not cooperatively.

Failure of eggs may be due to tidal flooding, pesticides, tourist disturbance and fishermen. Predators include raccoons taking nestlings, bald eagles and peregrine falcons.

There are five subspecies:

> *P.o. occidentalis*, Linnaeus, 1766.
> *P.o.carolinensis*, Gmelin, 1789.
> *P.o. californicus*, Ridgway, 1884.
> *P.o. murphyi*, Wetmore, 1945.
> *P.o. urinator*, Wetmore, 1945.

Peruvian Pelican: *Pelecanus thagus*

With similar plumage to the brown pelican, but noticeably larger (nearly twice the weight, at 7 kg (15.4 lb)) and longer, the Peruvian pelican and brown pelican are the only true marine pelican species.

The Peruvian pelican has dark plumage, dark facial skin and a white stripe from the top of its bill up to the crown, and then down the sides of the neck. It has long tufted feathers on the top of the head, pale upper wings and white streaks on its underside. The bill is yellow, with a reddish tip. Its blue striped pouch becomes brighter in the breeding season.[13]

Although both the brown pelican and the Peruvian pelican are found in the Americas, the Peruvian pelican keeps to the west coasts of Peru and Chile, restricted to the cold water of the Humboldt Current. 'The two overlap in the northern coast of Peru, which constitutes the brown pelican's southernmost point in its regular wintering ground along the Pacific.'[14] Pelicans can be aggressive towards other seabird species in the Humboldt Current region, because they are competing for nesting space or for food.[15] Peruvian pelicans need to rest and nest on flat areas on higher ground (as do imperial cormorants), because their long wings make it difficult to land or rest safely on cliffs, so there is competition for space between these two species.[16] Peruvian pelicans may also prey on cormorant nestlings, using them as a food source.[17]

Peruvian pelicans primarily plunge dive for food.[18] Their subcutaneous air sacs prevent them from diving more than a little way below the surface, which limits them.[19] Being resourceful, they steal fish brought to the surface by other seabirds.

References

INTRODUCTION

1 'Jerome Clark', https://en.wikipedia.org, accessed 12 February 2017. The description stuck after a UFO sighting by Kenneth Arnold in 1947. Sceptics said that the supposed UFO was just a group of pelicans..

1 BIOLOGICAL MARVELS

1 'They have a biography, not merely a biology.' Tom Regan, 'Animal Rights', in *Encyclopedia of Animal Rights and Animal Welfare*, ed. Marc Bekoff, 2nd edn, https://books.google.com.au, accessed 12 February 2017; for more information, see Chapter Four.
2 J. Bryan Nelson, *Bird Families of the World: Pelicans, Cormorants and their Relatives* (Oxford, 2005), p. 8.
3 The hamerkop and spoonbill were previously classified as belonging to the order Ciconiiformes but have been reclassified as Pelecaniformes.
4 Nelson, *Bird Families*, p. 6.
5 Ibid., p. 9.
6 Ibid.
7 Jeff Hecht, 'Pelican Fossil Poses Evolutionary Puzzle', www.newscientist.com, 22 June 2010. Scarcity of fossilized pelican beak finds has hampered knowledge about the early evolution.
8 Ibid.
9 Antoine Louchart, Nicolas Tourment and Julie Carrier, 'The Earliest Known Pelican Reveals 30 Million Years of Evolutionary

Stasis in Beak Morphology', *Journal of Ornithology*, CLII/1 (2011), pp. 15–20. Matt Walker, 'Oldest Prehistoric Pelican also had Big Beak', BBC Earth Watch, http://news.bbc.co.uk, 11 June 2010.

10   According to Dr Louchart, few creatures capable of flight have undergone so little change. He suggests that the bat may be the only other example (aside from the pelican), with a body shape much the same as it was 5 million years ago.

11   Museum of Zoology, University of Michigan, http://animaldiversity.org, accessed 12 February 2017. According to *Birdlife Australia*, the length of the Australian pelican's bill is between 40 and 50 cm, http://birdlife.org.au, accessed 12 February 2017; according to the *Guinness World Records*: 'The longest bill is that of the Australian pelican (*Pelecanus conspicillatus*); it is 34–47 cm (13–18½ in.) long.' www.guinnessworldrecords.com, accessed 12 February 2017.

12   Eberhard Fraas (1824–1897) was a professor in geology and paleontology. Fraas unearthed the fossil record of the *Pelicanus intermedius* in 1870: more information can be found in Richard *Lydekker's Catalogue of the Fossil Birds in the British Museum (Natural History)* (London, 1891), pp. 37–45, available at https://archive.org. A. H. Miller, an Australian avian paleontologist, found several fragments of a fossilized pelican, known as *Pelicanus tirarensis*, in the Lake Eyre Basin, south Australia, in 1966.

13   The Pliocene Epoch was 5.3–2.6 million years ago; other finds include a new species (*P. schreiben*) from the lower Pliocene in North Carolina. According to Olson, it represents a line with no living descendants. S. L. Olson, 'A New Species of Pelican (Aves: Pelecanidae) from the Lower Pliocene of North Carolina and Florida', *Proceedings of the Biological Society of Washington*, CXII (1999), pp. 503–9.

14   Nelson, *Bird Families*, p. 10. The Neolithic Age is difficult to date, but it is thought to have begun in the Middle East approx. 10,200 BCE (later elsewhere) and concluded 4,500–3,000 BCE.

15   See www.nzbirds.com, accessed 12 February 2017.

16   A summary of their argument is available at www.nzbirds.com, accessed 12 February 2017; a vagrant bird is a bird that has strayed or been blown off its usual range or migratory route. This term has been used to describe Trans-Tasman human relations as well.

17   See www.nzbirdsonline.org.nz, accessed 12 February 2017. One bird was still present on the Kaipara Harbour as late as March 2015.

18   According to Nelson, *Bird Families*, p. 270, an Australian Pelican can live up to sixty years in captivity. Wild Australian pelicans live for fifteen to 25 years.

19   Brett Westwood and Stephen Moss, *Natural Histories* (London, 2015), p. 142.

20   Pterosaurs were the first vertebrates to flap wings in order to fly. Pterosaurs lived alongside dinosaurs, but they were not dinosaurs themselves. According to researcher Xiaolin Wang, 'Pelicans and other modern birds with throat pouches are descended from dinosaurs, not pterosaurs, which were reptiles. Both *Ikrandraco* and pelicans may have separately evolved pouches and skimming flight. Quoted in Dan Vergano, 'Ancient Flying Reptile Ate Like a Toothy Pelican', *National Geographic*, http://news.nationalgeographic.com, 10 September 2014.

21   Xiaolin Wang et al., 'An Early Cretaceous Pterosaur with an Unusual Mandibular Crest from China and a Potential Novel Feeding Strategy', www.nature.com, 11 September 2014.

22   These specimens were primarily of *Rhamphorhynchus* and *Pterodactylus*; 'While some researchers have suggested a structure similar to that seen in pelicans . . . others have used the more neutral term "loose extensible skin", arguing that this gular structure might have helped them to swallow large prey items . . . None of the pterosaurs where they are thought to be present show the hook-shaped dentary process present in *Ikrandraco avatar*. It is possible that a gular structure potentially made of extensible skin was more developed in the new species.' Wang et al., 'An Early Cretaceous Pterosaur'.

23 It can also be a surname: Jaroslav Pelikan, Sterling Professor
    Emeritus of History at Yale; Ian Fraser and Jeannie Gray,
    Australian Bird Names: A Complete Guide (Collingwood, 2013), p. 62.
24 Ibid.
25 Diana Fernando, *Alchemy: An Illustrated A to Z* (London, 1998),
    p. 125.
26 John MacGregor, *The Rob Roy on the Jordan, Nile, Red Sea and
    Gennesareth, &c.: a canoe cruise in Palestine and Egypt, and the
    waters of Damascus* (London, 1869), p. 288, https://archive.org,
    accessed 12 February 2017.
27 John Giacon, 'Etymology of Yuwaalaraay Gamilaraay Bird Names',
    in *Lexical and Structural Etymology: Beyond Word Histories*, ed.
    Robert Mailhammer (online, 2013), p. 281, www.academia.edu,
    accessed 12 February 2017; see Chapter Two for the detailed story.
28 Jorge Luis Borges, *The Book of Imaginary Beings* (London, 2002),
    p. 114. I have been unable to find evidence for the naming of the
    genus *pelecanus* being linked to its hue, though colour does come
    into consideration when some species of pelican are named.
29 'When, somewhat more than thirty years ago, I first removed to
    Kentucky, Pelicans of this species were frequently seen by me on the
    sand bars of the Ohio and on the rockbound waters of the rapids
    of that majestic river . . . when a few years afterwards I established
    myself at Henderson, the White Pelicans were so abundant that
    I often killed several at a shot on a well-known sandbar which
    protects Canoe Creek Island. During those delightful days of my
    early manhood, how often have I watched them with delight!
    Methinks indeed, reader, those days have returned to me as if
    to enable me the better once more to read the scattered notes
    contained in my often-searched journals,' John James Audubon, *The
    American White Pelican*, in *The Audubon Reader*, ed. Richard Rhodes
    (New York, 2006), p. 520. I wonder if the pelicans were able to
    'wander free and unmolested' from Audubon's gun?
30 Maureen Lambourne, *The Art of Bird Illustration* (London, 2005),
    p. 41. Pelicans were among the water birds collected for the royal
    bird sanctuary, which still exists today.

31 Fraser and Gray, *Australian Bird Names*, p. 61.
32 Ibid., p. 57.
33 Taxonomy is the classification and naming of organisms in an ordered system that is intended to indicate natural relationships, especially evolutionary relationships.
34 Nelson, *Bird Families*, p. 5.
35 David Badke, 'Pelican', www.bestiary.ca, accessed 12 February 2017.
36 *Bestiaire divin* (the longest of the French bestiaries, written in 1210 or 1211), ibid.
37 Linnaeus is credited as the 'Father of Taxonomy'. His system for naming, ranking and classifying organisms is still in use today, though with many changes. The first part of the name identifies the genus to which the species belongs (*Pelecanus*), and the second part identifies the species within the genus (*onocrotalus*).
38 The weight range is 9 to 15 kg for the male, 3 to 9 kg for the female.
39 In 1828 Bruch proposed a system of trinomial nomenclature for species, in contrast to the binomial system of Linnaeus.
40 Fraser and Gray, *Australian Bird Names*, p. 62.
41 Stanley Clisby Arthur, 'The Emblematic Bird of Louisiana', http://penelope.uchicago.edu, accessed 12 February 2017. Bossu lived from 1720 to 1792.
42 Juan Ignacio Molina (1740–1829) was also known as Abbot Giovanni Ignazio. He was a Chilean Jesuit priest as well as a naturalist, historian and ornithologist.
43 Crispus, Murphyi, Carol, Phil, Californicus, the one who went journeying, Occidentalis, the 'western accident', the elderly one, known as urinator, and the one whose name we cannot remember, but may be foreign, always depicted with specs, our sweet Pélican à lunettes.
44 Martyn Kennedy et al., 'The Phylogenetic Relationships of the Extant Pelicans inferred from DNA Sequence *Data*', *Molecular Phylogenetics and Evolution*, LXVI (2013), www.sciencedirect.com, accessed 12 February 2017.
45 Merritt (1879–1972) was an American poet and humourist and editor of *The Tennessean*, a Nashville morning paper.

46 Fran Pickering, *The Illustrated Encyclopedia of Animals in Nature and Myth* (London, 2006), p. 97.

47 Walter Harter, *Birds in Fact and Legend* (New York, 1979), p. 74.

48 See 'Pelican Pouch Yoga, Phillip Island', 4 January 2013, http://petesflap.blogspot.com.

49 Colin Tudge, *The Secret Life of Birds* (London, 2009), p. 236. The technique resembles that of the blue whale: scooping up loads of water and fish, then straining off the water. After a catch of fish, the gular pouch can hold as much as 13 litres of water; Harter, *Birds in Fact and Legend*, p. 74.

50 Gregory McNamee, *Aelian's The Nature of Animals* (San Antonio, TX, 2011), p. 38. Another observation from Aelian: 'The pelican, I am told, does not like the quail, and the feeling is mutual.' Ibid.

51 Adele Nozedar, *The Secret Language of Birds: A Treasury of Myths, Folklore and Inspirational True Stories* (London, 2006), p. 328.

52 Badke, 'Pelican', www.bestiary.ca, accessed 12 February 2017.

53 Peter Tate, *Flights of Fancy: Birds in Myth, Legend and Superstition* (London, 2007), pp. 105–6.

54 From Bartholomaeus Anglicus, *De proprietatibus rerum*, book 12, trans. John Trevisa, www.bestiary.ca, accessed 12 February 2017.

55 Quoted in Nozedar, *The Secret Language of Birds*, p. 328.

56 Nelson, *Bird Families*, p. 261.

57 For the question concerning whether they compete with commercial fisheries, see Chapter Four.

58 Becky Bauer, 'Brown Pelicans – Myths and Facts Part 1', www.allatsea.net, 1 September 2005.

59 Ibid. Their bodies are filled with air sacs which cushion the force of impact when brown pelicans hit the water.

60 Nelson, *Bird Families*, p. 281.

61 Ibid.

62 Ibid.

63 Nelson, *Bird Families*, p. 109.

64 Ibid., p. 267. Material continues to be added to the nest, even after the young have hatched.

65 'Animal Species: Australian Pelican', www.australianmuseum.net.au, accessed 12 February 2017.

66 James Jennings, *The Family Cyclopaedia* (London, 1822), p. 907, available at https://books.google.com.au, accessed 17 February 2017.

67 David Gulpilil, *Stories of the Dreamtime*, compiled by Hugh Rule and Stuart Goodman (Sydney, 1979), p. 47.

68 Nelson, *Bird Families*, p. 127.

69 Scott Forbes, *A Natural History of Families* (Princeton, NJ, 2005), p. 11; 'Great white, Australian, and pink-backed pelicans rear only a single chick, whilst only a small proportion of American white and Dalmatian pelicans rear two, leaving only the brown and possibly the spot-billed that habitually rear two or more.' Nelson, *Bird Families*, p. 64.

70 Nelson, *Bird Families*, p. 33.

71 Forbes, *Natural History*, p. 11. A pelican rescue group in Australia was set up by twins, and is named Twinnies; but the name has nothing to do with siblicide!

72 Ibid., p. 113.

73 Ibid., p. 131. Fat levels may drop, feather replacement may be impaired and the extra workload could lead to a suppressed immune system, making the pelican more susceptible to infection and disease.

74 For more information, see Chapter Two.

75 Similar behaviour is also observed in human toddlers; think 'in the supermarket at 4 p.m. with a hungry two-year-old who didn't have an afternoon nap'.

76 Nelson, *Bird Families*, p. 253.

77 Ibid., p. 254.

78 Ibid., p. 124.

79 Cash and Evans's research was published in K. J. Cash and R. M. Evans, 'The Occurrence, Context and Functional Significance of Aggressive Begging Behaviours in Young American White Pelicans', *Behaviour*, CII (1987), pp. 119–28, and cited in Nelson, *Bird Families*, p. 124.

80 Nelson, *Bird Families*, p. 124.

81 Ibid., p. 125.

82 Ibid.

83 Pneumatic bird bones are a type of hollow bone filled with air that contain supporting trabecular (tissue in the form of a small beam, or rod); this combination of light weight and strength evolved to enable flight.

84 *Le Peuple Migrateur* (Winged Migration) is a documentary on the migratory patterns of birds, shot over three years on all seven continents.

85 Henri Weimerskirch et al., 'Energy Saving in Flight Formation', *Nature*, CDXIII (18 October 2001), pp. 697–8.

86 See 'Bird Flight Explained', http://news.bbc.co.uk, accessed 10 December 2018.

87 Mark Shields, 'Brown Pelican', www.birdsna.org, 15 July 2014.

88 Nelson, *Bird Families*, p. 244.

89 'American White Pelican', www.borealbirds.org, accessed 26 March 2018.

90 Javier Gutierrez Illan, Guiming Wang, Fred L. Cunningham and D. Tommy King, 'Seasonal Effects of Wind Conditions on Migration Patterns of Soaring American White Pelican', http://journals.plos.org, accessed 17 March 2018.

91 Mean hourly speed during the migration was 39.53 km per hour (spring) and 33.16 km per hour (autumn); ibid.

92 Ibid.

93 Ibid. Habitat protection along migration corridors must also continue.

2 FROM CREATION MYTHS TO CONCRETE BEHEMOTH

1 Examples include the raven, for the first peoples in North America, and the dove for the three Abrahamic faiths (Judaism, Christianity and Islam).

2 Adele Nozedar, *The Secret Language of Birds: A Treasury of Myths, Folklore and Inspirational True Stories* (London, 2006), p. 328. This act is similar to a genre of creation stories known as 'diver myths',

in which a creature dives into primordial waters to collect mud, which is then fashioned into land. Barbara Allen, *Animals in Religion* (London 2016), pp. 33–4.

3 According to Australian indigenous belief, the Dreamtime was the ancient time when all things were created by sacred ancestors. All life as it is found today (animal, human and plant), and its interrelatedness, can be traced back to the Dreamtime.

4 Nozedar, *The Secret Language of Birds*, p. 329.

5 Michael J. Connolly, 'Goolay-Yali the Pelican', www.kullillaart. com.au, accessed 12 February 2017.

6 Ibid.

7 Roland Robinson, *Aboriginal Myths and Legends* (Melbourne, 1966), p. 151.

8 A. W. Reed, *Aboriginal Myths: Tales of the Dreamtime* (Frenchs Forest, 1978), p. 37.

9 Ibid., p. 39.

10 David Gulpilil, *Stories of the Dreamtime*, compiled by Hugh Rule and Stuart Goodman (Sydney, 1979), pp. 47–53. In another version, Booran the pelican seeks revenge, paints himself with white clay, but meets a bigger black pelican, who kills him. 'Ever since that day . . . the plumage of the Pelicans has been a mixture of black and white.' A. W. Reed, *More Aboriginal Stories of Australia* (Sydney, 1980), p. 17.

11 Philomene Corrigal, 'Humility', www.facebook.com/ UnifiKshuNaShun, accessed 26 February 2017.

12 The word 'Henet' is similar to the ancient Egyptian word for queen ('Hent') and to the word for priestess (Henut).

13 *Ancient Egyptian Bestiary*, www.reshafim.org.il, accessed 12 February 2017.

14 This is number 226 in the utterances of the Unas text, www.pyramidtextsonline.com, accessed 26 November 2018.

15 Joann Fletcher, 'Mummies Around the World', www.bbc.co.uk, 17 February 2011.

16 Charles C. Mann, *1491: New Revelations of the Americas Before Columbus* (New York, 2011), p. 214.

17 James W. Reid, *Magic Feathers: Textile Art from Ancient Peru* (London, 2005).

18 Kira Kathaleen Jones, 'Pelican', https://scholarblogs.emory.edu, accessed 12 February 2017.

19 The state of the friezes has led some to question their authenticity. The friezes are mostly complete, and appear to be in perfect condition, so are they replicas?

20 Michael Bright, *Beasts of the Field: The Revealing Natural History of Animals in the Bible* (London, 2006), p. 226.

21 Ibid., p. 227.

22 Roy Pinney, *The Animals in the Bible: The Identity and Natural History of All the Animals Mentioned in the Bible* (Philadelphia, PA, 1964), p. 140.

23 The Pelican Lectern in Norwich Cathedral was saved from the destruction during the Reformation and was later discovered in the bishop's garden, where it had been buried for years.

24 The Greek *physiologus*, 'a discourse on nature', is the title of a work probably of the third century, but was thought to be the name of the author, 'the naturalist'. J.C.J. Metford, *Dictionary of Christian Lore and Legend* (London, 1983), p. 199. There is disagreement among scholars concerning its date; some say the work is from the mid-second century CE, which is the date I have included in the Timeline.

25 William Saunders, 'The Symbolism of the Pelican', *Arlington Catholic Herald*, www.catholiceducation.org, accessed 12 February 2017.

26 Walter Harter, *Birds in Fact and Legend* (New York, 1979), p. 76.

27 J. Bryan Nelson, *Bird Families of the World: Pelicans, Cormorants and their Relatives* (Oxford, 2005), p. 95. Nelson notes that pelicans do treat their young harshly.

28 Udo Becker, ed., *The Element Encyclopedia of Symbols* (Longmead, 1986), p. 230.

29 Mike Klug, 'Sacred Symbols: The Pelican', https://mikejklug.wordpress.com, 27 August 2014.

30   Leonardo da Vinci, 'The Pelican', *Fables of Leonardo da Vinci*, interpreted and transcribed by Bruno Nardini (London, 1973), p. 56. In another version, which makes the Christian connection clearer, the snake blows venom on the bird's young and kills them. When their mother sees that they are dead, 'she looks up to a cloud and flies there; and striking her side with her wings till blood streams, she lets the drops fall . . . on the young ones and they come to life again.' Edward A. Armstrong, *The Life and Lore of the Bird: In Nature, Art, Myth, and Literature* (New York, 1975), p. 84.

31   John Sparks, *The Discovery of Animal Behaviour* (London, 1982), p. 57.

32   Maureen Lambourne, *The Art of Bird Illustration* (London, 2005), p. 36.

33   George Towner, 'The Case of the Pious Pelican', www.ecphorizer. com, accessed 12 February 2017.

34   J. C. Galton, 'Does the Pelican Feed Its Young With Its Blood?', *Notes and Queries*, IV (1869), pp. 361–2, available at http:// penelope.uchicago.edu, accessed 12 February 2017. White notes that 'It was suggested by a Mr A. D. Bartlett (*Proc. Zool. Soc.* 1869, p. 146) that the pelican of piety was really a flamingo, which does eject "a curious bloody secretion from its mouth." But probably the ordinary ejection of food from the pelican's pouch may account for the legend.' T. H. White, ed., *The Book of Beasts: Being a Translation from a Latin Bestiary of the Twelfth Century* (New York, 1984), p. 133.

35   *The Spirit of Gertrude* (London, 1866), p. 93, available at www.catholickingdom.com, accessed 12 February 2017.

36   Saunders, 'The Symbolism of the Pelican'.

37   Ibid.

38   Rees Prichard, 'Christ is All in All', https://allpoetry.com, verses 32 and 33, accessed 12 February 2017.

39   *Catholic Encyclopaedia* (1913), quoted in 'Adoro te devote', https://en.wikipedia.org, accessed 12 February 2017.

40   Charles Price, 'Proposed Texts', *The Living Church*, CLXXXV (1992), p. 201, https://books.google.com.au, accessed 12 February 2017.

41 Raymond F. Glover, *The Hymnal 1982 Companion*, vol. I
(New York, 1990), p. 612, https://books.google.com.au,
accessed 12 February 2017.

42 The Pelican Foundation is also the name of a charity in
southeast Michigan. Other groups come under the banner
of 'pelican' to speak of their work, including The Society
for Creative Anachronism, an international organization
dedicated to researching and recreating the arts and skills
of pre-seventeenth-century Europe. One of their Orders is
known as the 'Pelican'.

43 David Badke, 'Pelican', www.bestiary.ca, accessed 12 February
2017.

44 Ivan Clutterbuck, *The Pelican in the Wilderness* (Leominster, 2008).

45 Robert W. Griggs, *A Pelican of the Wilderness* (Eugene, OR, 2014).

46 Isabel Colegate, *A Pelican in the Wilderness: Hermits, Solitaries and
Recluses* (New York, 2002); Thomas Traherne (1637–1674), ibid.,
p. 262.

47 Nelson, *Bird Families*, pp. 119 and 247. 'In the Tel Aviv zoo,
pelicans didn't breed until surrounded by mirrors.'

48 Zoe Craig, 'Everything You Need to Know about the St James's
Park Pelicans', *Londonist*, https://londonist.com, accessed 4
December 2018. For the House of Lords proceedings, Wednesday
20 December 1995, please see https://publications.parliament.uk.

49 'Early Scottish Rite Tortoise Shell Snuff Box', www.
phoenixmasonry.org, accessed 12 February 2017. Resurrection
is a doctrine held in the Masonic Degree, with its emphasis
on the old Temple being destroyed and a new Temple
springing forth.

50 Conor Moran, 'The Role of the Pelican Inside and Outside the
Walls of the Scottish Rite', http://aasrnys.blogspot.com.au,
17 March 2010.

51 Ibid. I have been unable to confirm this; when I have
examined illustrations of pelicans in bestiaries and other art
work, sometimes the wound is on the right-hand side, on others,
in the middle, or to the left. Perhaps the point is that emulating

legendary qualities of the pelican may help one on the path of self-sacrifice.

52 Becker, ed., *The Element Encyclopedia of Symbols*, p. 23.

53 Nozedar, *The Secret Language of Birds*, p. 222.

54 Ted Andrews, *Animal Speak: The Spiritual and Magical Powers of Creatures Great and Small* (St Paul, MN, 1993), p. 183.

55 Ibid.

56 Ibid.

57 'The Story of Sanquin',www.sanquin.nl, accessed 1 December 2018.

58 Jenny Stratford, *Richard II and the English Royal Treasure* (Woodbridge, 2012), p. 53, available at https://books.google.com.au, accessed 12 February 2017.

59 Judith M. Bennett, *Queens, Whores and Maidens: Women in Chaucer's England*, Hayes Robinson Lecture Series No. 6 (2002), p. 13, presented at Royal Holloway, University of London, 5 March 2002, www.royalholloway.ac.uk, accessed 12 February 2017.

60 Daniel Hahn, *The Tower Menagerie* (London, 2003), p. 136. John Evelyn was also a founding member of the Royal Society.

61 Ibid., p. 208.

62 Alistair Kerr, 'How the St James's Park Pelicans Sparked a Cold War Stand-off between Russia and the USA', *Country Life*, 4 January 2018, www.countrylife.co.uk.

63 George Towner, 'The Case of the Pious Pelican'.

64 John Vinycomb, *Fictitious and Symbolic Creatures in Art* (London, 1909), available at www.sacred-texts.com, accessed 12 February 2017.

65 'The Louisiana State Flag', www.netstate.com, accessed 12 February 2017.

66 Bruce Atherton, *Pelican Pat* (Melbourne, 2006). Birds do have knees, but we usually cannot see them because they are situated on the upper leg, hidden by feathers of the wings and body.

67 'About Us', www.pelicancancer.org, accessed 12 February 2017.

68 *Brewer's Dictionary of Phrase and Fable*, revised by Adrian Room (London, 2001), p. 894. The acronym was assimilated from the

original Pelicon Crossing – PEdestrian LIght CONtrolled Crossing. *Pelican Crossing* is also the title of a book by Hilary Cotterill that narrates the ups and downs in her nursing training.

69 'World's Largest Pelican', www.roadsideamerica.com/story/2134, accessed 12 February 2017.

70 The emblem was later changed to the boronia, a plant.

71 Jonathan Clements and Helen McCarthy, *The Anime Encyclopedia, 3rd Revised Edition: A Century of Japanese Animation* (Albany, NY, 2015), https://books.google.com.au.

72 Herman von Kradenburg, 'Pelican 16 Down: The Sad Story of a SAAF Shackleton (Part 3 of Shackleton MR3 in SAAF Service)', http://aircraftnut.blogspot.com.au, 14 October 2013.

73 'Boeing Pelican', http://forum.worldofwarplanes.eu, accessed 12 February 2017.

74 Matthew Brown, 'Pelecanus Onocrotalus, Great White Pelican', http://blogs.bu.edu, accessed 12 February 2017. Interestingly, this is the blog of the course module 'Bio-aerial Locomotion', part of Introduction to Engineering at Boston University's College of Engineering.

75 Perhaps renaming the ship was also to appease his patron, for during the voyage one of the gentlemen on board, Thomas Doughty (who was also Hatton's personal secretary), had been put on trial for sedition and mutiny, found guilty, then executed. The ship's log entry also included the change of name to the *Golden Hind*: 'Jun 20 1578: Thomas Doughty tried and executed for mutiny . . . Pelican renamed Golden Hind.' Linda Alchin, 'The Golden Hind Ship', www.elizabethan-era.org.uk, accessed 12 February 2017. Drake's ship, the *Pelican*, was featured on the British halfpenny.

76 On 20 July 1991, on the ABC radio news: 'A chihuahua dog has come to a sticky end on North Stradbroke Island, off Brisbane. The chihuahua, owned by tourists, became excited when a pelican landed at the end of a jetty on the island. The chihuahua raced up to the pelican and barked repeatedly. Onlookers were stunned when, after the dog refused to back off, the bird opened its bill

and grabbed the chihuahua.' Bill Scott, *Pelicans and Chihuahuas and Other Urban Legends* (St Lucia, 1996), p. 7.

77 *Sydney Morning Herald*, 29 July 1991, quoted ibid., pp. 7–8.

78 Ross Bilton, 'John Ayliffe is the Pelican Man of Kangaroo Island', *The Australian*, 13–14 February 2016, p. 7.

3 GRACING THE PAGE AND SCREEN

1 Robert McCrum, 'What Would Allen Lane Make of Amazon?', www.theguardian.com, 27 September 2013.

2 Michael Bright, *Beasts of the Field: The Revealing Natural History of Animals in the Bible* (London, 2006), p. 226. Jugge is cited as the inventor of the footnote! Thank you, Mr Jugge.

3 Lindsey Hager, 'The Myth Behind the Pelicans' Large Beak', https://hagerl3.wordpress.com, 5 March 2014.

4 Richard de Fournival, *Master Richard's Bestiary of Love and Response*, trans. Jeanette Beer (Berkeley, CA, 1986), pp. 19–20.

5 Susan Hyman, *Edward Lear's Birds* (New York, 1980), pp. 45, 49.

6 Peter Levi, *Edward Lear: A Biography* (London, 1995), p. 119.

7 Edward Lear, 'Avlóna', *Journals of a Landscape Painter in Greece and Albania, &c.* (London, 1851) quoted in Hyman, *Edward Lear's Birds*, p. 65.

8 Quoted in Vivien Noakes, *Edward Lear* (London, 1985), p. 176.

9 Edward Lear, 'The Pelican Chorus', *The Complete Nonsense of Edward Lear*, ed. and intro. Holbrook Jackson (London, 1984), pp. 232–3. Music is provided in the book; arrangement for the piano by Professor Pomé of Sanremo, Italy.

10 Levi, *Edward Lear: A Biography*, p. 119.

11 Susan Chitty, *That Singular Person Called Lear* (Stroud, 2007), p. 41.

12 Ibid.

13 Hyman, *Edward Lear's Birds*, p. 88.

14 John Ciardi, 'The Reason for the Pelican', 1959, www.nowaterriver.com, accessed 12 February 2017. Ciardi (1916–1986) wrote for adults, then branched into children's literature in 1959 with the publication of a book entitled *The Reason for the Pelican.*

15 Robert Desnos, 'Le Pélican', www.poesie.net, accessed
    12 February 2017.

16 See primary school children reciting the poem: www.youtube.
    com, accessed 20 March 2018. It is one of the online teaching
    tools used at https://quizlet.com, accessed 20 March 2018.

17 Fleur Adcock and Jacqueline Simms, eds, *The Oxford Book of
    Creatures* (Oxford, 1997), p. 3.

18 George MacDonald, *The Seaboard Parish,* in *The Parish Papers
    (Three Novels in One)* (Colorado Springs, CO, 1985), p. 223.
    In another of his works, *Robert Falconer* (1868), MacDonald,
    a Congregational minister as well as a writer, and therefore very
    familiar with the Psalms, draws on Psalm 102:6: 'Whaur hae ye
    been this mony a day, like a pelican o' the wilderness?' George
    MacDonald, *Robert Falconer*, vol. I, https://books.google.com.au,
    accessed 12 February 2017.

19 James Joyce, *Ulysses,* Chapter 14, www.online-literature.com,
    accessed 12 February 2017.

20 Hans Christian Anderson, 'The Garden of Paradise', in *The Complete
    Illustrated Stories of Hans Christian Anderson* (London, 1983,
    from an 1889 edition), p. 648.

21 Wayne Talbot and Greg McKee, *The Great Wungle Bungle Aerial
    Expedition* (Richmond, Victoria, 1992).

22 Narelle Oliver, *The Best Beak in Boonaroo Bay* (Port Melbourne,
    Victoria, 1995).

23 Graeme Base, *My Grandma Lived in Gooligulch* (Camberwell,
    Victoria, 1983), p. 21.

24 Julie Watts, *The Art of Graeme Base* (Melbourne, 2009), p. 35.

25 Although Thiele wrote over a hundred books, was an educator
    and a passionate conservationist, there was little media interest
    in his death on 4 September 2006. Reports of Thiele's death
    were eclipsed by the death of another on the same day: wild-life
    wrangler Steve Irwin. The true environmentalist was overlooked.

26 Sonia Harford, '*Storm Boy* illustrator Robert Ingpen on bringing
    Colin Thiele's classic book to life, and seeing it adapted for the
    stage', www.smh.com.au, 4 May 2015.

27 Colin Thiele, *Storm Boy*, illus. Robert Ingpen (Sydney, 1992), p. 21.
28 Ibid., p. 56.
29 Ibid., p. 61.
30 Michaela Boland, 'Kicking up a Storm with Colin Thiele's Classic Boy's Tale', www.theaustralian.com.au, 3 August 2013.
31 'Storm Boy Pelican Dies', www.abc.net.au, September 2009.
32 Boland, 'Kicking up a Storm'. A new film of *Storm Boy* is due for release in 2019. In this movie version, Storm Boy has grown up, and is now a grandfather, guiding his granddaughter through troubled times. He tells her about Mr Percival, the orphaned pelican he rescued when he was a child, and the influence this incident had on his life.
33 Michael Page and Robert Ingpen, *Encyclopedia of Things That Never Were* (London, 1997), p. 20.
34 Robert Ingpen, *The Voyage of the Poppykettle* (Adelaide, 1980), p. 23.
35 Robert Ingpen, *The Unchosen Land* (Adelaide, 1981), p. 14.
36 Michael Lawrence, *The Poppykettle Papers*, illus. Robert Ingpen (London, 1999), verso page.
37 Ibid., p. 21.
38 Ibid., p. 112.
39 'Finding Nemo: Goofs', www.imdb.com, accessed 12 February 2017.
40 'The Adventures of Paddy the Pelican', https://en.wikipedia.org, accessed 12 February 2017.

4 THE END OF THE GOLDEN AGE

1 Diary entry, 4 April 1802, Matthew Flinders speaking about Kangaroo Island, an island he had named in March that year. Matthew Flinders, *A Voyage to Terra Australis: Undertaken for the purpose of completing the discovery of that vast country*, vol. 1 (London, 1814, Australiana Facsimile Editions No. 37, Adelaide, 1966), pp. 183–4.
2 Edward A. Armstrong, *The Life and Lore of the Bird: In Nature, Art, Myth, and Literature* (New York, 1975), p. 196. The arguments

were a resounding success: on 1 March 1872 Yellowstone Park was established, the first of many such parks.

3 David Kenagy, 'Kenai National Wildlife Refuge Volunteers Wear Many Hats, but None with Plumes', http://peninsulaclarion.com, 24 May 2002.

4 Poem by Athelinus in *The Mirror of Literature, Amusement, and Instruction, Containing Original Papers*, vol. VIII (London, 1845), p. 312, available at https://books.google.com.au, accessed 12 February 2017. There is some disagreement regarding the species of bird; some say it was a pelican, others that it was a bittern.

5 Kenagy, 'Kenai National Wildlife Refuge Volunteers Wear Many Hats, but None with Plumes',

6 Duff Hart-Davis, *Audubon's Elephant* (London, 2004), p. 163.

7 John J. Audubon to John Bachman, 'The weather was fair and the sea was smooth', letter of 6 April 1837 written aboard the U.S. Revenue Cutter *Campbell*, Island of Barataria, Grande Terre, in John J. Audubon, *The Audubon Reader*, ed. Richard Rhodes (New York, 2006), p. 525.

8 'He was always searching not only for new species, but for small variations within species – between male and female, between juvenile and mature – and at the same time seeking to defray his own expenses by collecting skins which he could sell to European museums.' Hart-Davis, *Audubon's Elephant*, p. 164.

9 John MacGregor, *The Rob Roy on the Jordan, Nile, Red Sea and Gennesareth, &c.: a canoe cruise in Palestine and Egypt, and the waters of Damascus* (London, 1869), pp. 287–8, available at https://archive.org, accessed 12 February 2017.

10 Peter Tate, *Flights of Fancy: Birds in Myth, Legend and Superstition* (London, 2007), p. 106.

11 John J. Audubon, 'American White Pelican', *Birds of America*, available at www.audubon.org, accessed 12 February 2017.

12 J. Bryan Nelson, *Bird Families of the World: Pelicans, Cormorants and their Relatives* (Oxford, 2005), p. 95.

13   John J. Audubon, 'Brown Pelican', *Birds of America*, available
     at www.audubon.org, accessed 12 February 2017.
14   Antoine-Simon Le Page du Pratz (1695–1775) was an
     ethnographer, historian and naturalist, best known for his
     Histoire de la Louisiane; Stanley Clisby Arthur, 'The Emblematic
     Bird of Louisiana', http://penelope.uchicago.edu, accessed
     on 12 February 2017.
15   James Jennings, *The Family Cyclopaedia* (London, 1822),
     p. 907, available at https://books.google.com.au, accessed
     17 February 2017.
16   P. L. Simmonds, 'Notes on Some Animal Oils and Fats
     used in Pharmacy', *The Bulletin of Pharmacy*, VII (1893), p. 297,
     https://archive.org, accessed 12 February 2017.
17   Nelson, *Bird Families*, p. 95. Pelican guano is inferior to that
     of Guanay cormorants and Peruvian boobies.
18   Nayambayar Batbayar et al., 'Conservation of the Critically
     Endangered East Asian population of Dalmatian Pelican
     Pelecanus crispus in Western Mongolia', *Birding Asia*, VII (2007),
     pp. 68–74, http://orientalbirdclub.org, accessed 12 February 2017.
19   John May and Michael Marten, *The Book of Beasts* (Feltham, 1982),
     p. 33.
20   Lawrence Pope, *The Teeth of Beasts* (Brisbane, 2012), p. 458.
21   Ibid., p. 463. Two years on, this measure seems to have been
     effective for there have been no further reports that pelicans
     have been killed on the lines.
22   A. J. Crivelli et al., 'Conservation and Management of Pelicans
     Nesting in the Palearctic', in *Conservation of Migratory Birds*,
     ed. T. Salathé (Cambridge, 1991), pp. 137–52.
23   Nelson, *Bird Families*, p. 95.
24   Becky Bauer, 'Brown Pelicans: Myths and Facts Part 1',
     www.allatsea.net, 1 September 2005.
25   Ibid.
26   Justine Frazier and Jonathan Atkins, 'If Pelicans could Cry –
     the Passing of the Pelican Man', www.abc.net.au, accessed
     11 November 2018.

27 Ibid. Marny Bonner, his partner, said: 'We could open a tackle shop on the hooks and lines and sinkers and traces and lures that we have extracted from the pelicans we've caught to date, easily.'

28 'Pelicans most at risk from fishing tackle injuries', http://phys. org, 17 July 2014. The results of the study were published in E. R. Carapetis et al., 'Recreational Fishing-related Injuries to Australian Pelicans (*Pelecanus Conspicillatus*) and Other Seabirds in a South Australian Estuarine and River Area', *International Journal of Veterinary Health Science & Research*, 11 (2014), available at http:// scidoc.org.

29 Nelson, *Bird Families*, p. 281.

30 Ibid.

31 Alfredo Bergazo, 'The Brown and Peruvian Pelicans', www.10000birds.com, 8 October 2015.

32 Ted Williams, 'Brown Pelicans: A Test Case For the Endangered Species Act', *Yale Environment 360*, http://e360.yale.edu, 12 May 2014.

33 Ibid.

34 Ibid.

35 Ibid.

36 Ibid.

37 Ibid.

38 Ibid.

39 L. Bunkley-Williams et al., 'The South American sailfin Armored Catfish, *Liposarcus multiradiatus* (Hancock), a New Exotic Established in Puerto Rican Fresh Waters', *Caribbean Journal of Science*, xxx (1994), pp. 90–94, available at biology.uprm.edu, accessed 17 April 2018.

40 Ibid., p. 90.

41 'Pelican Island: History', www.fws.gov, 14 October 2015. Overfishing is not a new ecological problem.

42 Nelson states that hundreds of pelicans have been shot in Israel in recent years. Nelson, *Bird Families*, p. 100.

43 Ruth Schuster, 'Israeli Animal Hospital Rescues Pelican With 110 Shotgun Pellets', www.haaretz.com, 31 December 2013.

44   Zafrir Rinat, 'Kibbutzim Sue Nature Authority for Letting Pelicans Eat Their Fish', www.haaretz.com, accessed 31 August 2015.

45   Ibid.

46   Schuster, 'Israeli Animal Hospital'.

47   Ibid.

48   'The Three White Pelicans of St James's Park', https://satnavandcider.wordpress.com, 11 October 2012.

49   Ibid.

50   Frank Messina, 'Pelican Mutilations Return with Vengeance', http://articles.latimes.com, 20 September 1992. This gives us some insight into some of the more 'routine' cruelty that has been inflicted on the federally protected pelican.

51   Ibid.

52   Reuters and Mia de Graaf, 'Serial Throat-slasher Kills 10 pelicans in South Florida a Year after Identical Incident that Wiped Out Dozens of Birds', www.dailymail.co.uk, 25 January 2015.

53   Nelson, *Bird Families*, p. 100.

54   Ibid.

55   Aleisha Orr, 'Pelican Used for "Target Practice"', www.watoday.com.au, 28 May 2013.

56   Reconstructive surgery on a pelican's beak was performed in March 1982 on Paul, an American White Pelican which resided in a zoo in Maryland. Paul's artificial beak was fashioned by veterinarians at the National Zoo in Washington, DC. Although Paul's artificial beak did not include a horn, which is a vital component of normal courting display, zoo officials said the fibreglass bill would function as a normal bill in every other way, allowing him to feed and preen. 'Paul the Pelican gets Fiberglass Bill', www.upi.com, 23 March 1982. There is no online record of the success or otherwise of the procedure.

57   'Veterinarians using Pins, Screws and Glue Attached an Artificial Beak to a Maimed Pelican', www.upi.com, 24 October 1982.

58   Eddie Krassenstein, 'Beautiful White Pelican Receives New Beak Thanks to 3D Printing Technology', www.embodi3d.com, accessed 10 August 2015.

59 Ibid.

60 It was also reported that they would redesign Pierre's head to make him 'less scary'. 'Pierre the Pelican Getting New Look', www.espn.com.au, 12 February 2014.

61 Ben Golliver, 'Pelicans' Mascot Pierre to Get New Look After 'Undergoing Reconstructive Beak Surgery', www.si.com, 11 February 2014. See also 'Pierre The Pelican', http://knowyourmeme.com, accessed 12 February 2017.

62 Louis Sahagun, 'California Brown Pelicans Found Frail and Far from Home', http://articles.latimes.com, 6 January 2009.

63 L. J. Davenport, *Nature Journal* (Tuscaloosa, AL, 2010), p. 140, available at https://books.google.com.au, accessed 12 February 2017.

64 Nelson, *Bird Families*, p. 62.

65 Ibid., p. 100.

66 Joseph K. Gaydos , 'Deadly Diatoms: The Latest on Harmful Algal Blooms', Proceedings of the 2012 North American Veterinary Conference, Orlando, Florida, p. 1, www.researchgate.net, accessed 12 February 2017.

67 Bauer, 'Brown Pelicans'.

68 Gaydos, 'Deadly Diatoms: The Latest on Harmful Algal Blooms'.

69 A. S. Beltran et al., 'Seabird Mortality at Cabo San Lucas, Mexico: Evidence that Toxic Diatom Blooms are Spreading', *Toxicon*, xxxv (1997), pp. 447–53, cited in Nelson, *Bird Families*, p. 128, no page number given.

70 Mary-Louise Vince, 'Mystery Solved: Hundreds of Pelican Deaths Explained', www.abc.net.au, 19 September 2016.

71 Hannah Nevins et al., 'Summary of Unusual Stranding Events Affecting Brown Pelican along the U.S. Pacific Coast during Two Winters, 2008–2009 and 2009–2010', www.academia.edu, p. 4, accessed 12 February 2017.

72 Ibid.

73 Center for Biological Diversity, 'A Deadly Toll: The Gulf Oil Spill and the Unfolding Wildlife Disaster', www.biologicaldiversity.org, April 2011, accessed 12 February 2017. While not on the

same scale, the Refugio oil spill at Santa Barbara, California on 19 May 2015 led to 3,400 barrels of crude oil leaking into the water, north of Refugio State Beach. The oil slick reached four marine protected areas, each being of ecological significance. Hundreds of animals along the coast died. Of the 69 animals freed after being cleaned and nursed back to health, twelve were adult brown pelicans. At least five pelicans have been fitted with solar-powered satellite tracking devices which will help scientists study them when they return to the wild. They were later released on 12 June 2015 at Goleta Beach. See 'Tracking Oiled Pelicans After the Refugio Oil Spill', www.evotis.org, and 'Brown Pelicans Released Following Refugio Oil Spill', http://blogs.ucdavis.edu, both accessed 2 December 2018.

74 Other sources suggest larger volumes. The Center of Biological Diversity reported in 2011 that the 'catastrophe . . . spilled 205.8 million gallons of oil and 225,000 tons of methane into the Gulf of Mexico. Approximately 25 percent of the oil was recovered, leaving more than 154 million gallons of oil at sea.' 'A Deadly Toll', p. 2.

75 Pallardy, 'Deepwater Horizon'.

76 Ibid.

77 Ibid.

78 Laura Tangley, 'Oil Spill Hammers Brown Pelicans', www.nwf.org, 15 September 2010.

79 Ibid.

80 Center for Biological Diversity, 'A Deadly Toll', p. 3. Based on government figures, news reports and scientific articles, the Center for Biological Diversity estimated that 'the oil spill had likely harmed or killed approximately 82,000 birds of 102 species, approximately 6,165 sea turtles, and up to 25,900 marine mammals'.

81 Ibid., p. 3.

82 Martha Harbison, 'Research Calculates the Cost of the Largest Oil Spill in U.S. History', www.audubon.org, 6 May 2014. The study by researchers Christopher Haney, Harold Geiger and Jeffrey

Short estimated that an additional 200,000 off-shore birds died during the acute phase.

83   Lacey McCormick, 'Five Years and Counting: Gulf Wildlife in the Aftermath of the Deepwater Horizon Disaster', www.nwf.org, 30 March 2015, p. 9.

84   Ibid., p. 10.

85   'The Gulf Oil Spill and the Unfolding Wildlife Disaster', Center of Biological Diversity Report, April 2011, www.biologicaldiversity. org, accessed 12 February 2017.

86   'Evidence of Oil Spill 2 Years Ago in Birds', www.upi.com, accessed 11 November 2018. Minnesota contains the largest lakeside colony of American white pelicans in North America.

87   Ibid.

88   Craig Welch, 'Is Gulf Oil Spill's Damage Over or Still Unfolding?', http://news.nationalgeographic.com, 14 April 2015. Were the booms a success? Back in 2010, Mike Parr, vice president of the American Bird Conservancy, after having completed a six-day visit to the Gulf, said: 'From what we observed, the boom they're using to keep oil out of critical areas is hopeless. It doesn't stand up to even moderate weather.' Quoted in Tangley, 'Oil Spill Hammers Brown Pelicans'.

89   'Krewe of Dead Pelicans', www.multispecies-salon.org, accessed 12 February 2017.

90   Natalie Richards: 'Black Pelican baffles Bird Enthusiasts', *The West Australian*, https://thewest.com.au, 18 May 2016.

91   The IUCN (International Union for Conservation of Nature), established in 1948, is an international organization working in the field of conservation. The IUCN Red List of Threatened Species (IUCN Red List) is the world's most comprehensive inventory of the global conservation status of species; 'Pelecanus crispus', http://www.iucnredlist.org, accessed 12 February 2017.

92   Batbayar et al., 'Conservation', p. 72. 'Muskrats damage reed beds, by eating reed shoots and roots. This kills the tussocks, causing holes and burrows, making them unsuitable for nesting purposes.'

93 Ibid., p. 71. 'Over 30,000 head of livestock from Chandmani,
    Mankhan, Buyant and Jargalant districts graze grass on the
    islands and inhabit the area until the end of spring in the Tsagaan
    Gol area in the southern part of Khar Us Lake, and in the Nariin
    Gol area in the north-western part of the same, where pelicans
    regularly summer and attempt to breed.'

94 Ibid.

95 Luiza Ilie, 'Romania Confirms Bird Flu in Dead Pelicans',
    www.reuters.com, 27 March 2015.

96 Not long after, H5N1 was detected in two dead pelicans in
    Bulgaria, in a nature reserve close to Romania. There were also
    cases in Russia (in April 2015) and Kazakhstan (in May 2015). This
    was not the first instance of avian flu affecting pelicans: in 2009
    H3N6 avian influenza virus was detected in a wild white pelican
    in Zambia, and in 2011 two pelicans were killed in Tokiwa Park in
    Japan to prevent a possible outbreak due to an ill mandarin duck.

97 The President of the Croatian Ornithological Society, Dragan
    Radovic, hopes that this pelican is not a one-off. 'We mustn't
    forget that the name in English is "Dalmatian Pelican" and that
    it is famous throughout the world by that name. [The bird]
    received its name from the Neretva's now extinct population.'
    Croatian Times, 24 March 2011, quoted at Loren Coleman,
    'Croatian Dalmatian Pelican Spotted', www.cryptomundo.com,
    24 March 2011.

98 'Trapped Dalmatian Pelicans Hand-fed in Frozen Caspian Sea',
    www.bbc.com, 21 February 2012.

99 'Pelecanus philippensis', www.iucnredlist.org, accessed
    12 February 2017.

100 Ibid.; Nelson, p. 242.

101 Ibid., p. 243.

102 Bridget Stutchbury, *The Private Lives of Birds* (New York, 2010),
    p. 181.

103 '*Pelecanus thagus*', www.iucnredlist.org, accessed 12 February 2017.

104 'Peru Examines Death of More than 500 Pelicans', www.bbc.com,
    30 April 2012.

105 Albert Schweitzer, *The Story of My Pelican* (London, 1964), p. 10.
106 Charles R. Joy, *The Animal World of Albert Schweitzer* (Hopewell, NJ, 1950), p. 21. Joy's book has this dedication: 'To Monsieur le Pélican, faithful guardian of Albert Schweitzer.'
107 Ibid., p. 25.
108 'Pelican Dreams', www.pelicanmedia.org, accessed 12 February 2017. Photo stills are on the site as well.
109 Hugh Powell, 'Pelican Dream s: The Cornell Lab Movie Review', www.allaboutbirds.org, 5 November 2014.
110 Caption below a photograph of a pelican, facing the contents page, in Joy, *Animal World of Albert Schweitzer*.

CONCLUSION

1 Emerald Pellot, 'Pelican Abandoned By Flock Learns To Fish From His Human', www.littlethings.com. The video is available at Ruby Valentine, 'Incredible Footage Of A Pelican Learning To Fly!', www.littlethings.com, accessed 12 February 2017.
2 'Theodore Roosevelt Quotes', www.nps.gov, accessed 12 February 2017.

APPENDIX

1 Weight range is 9–15 kg (20–33 lb) for the male, 3–9 kg (6.6–20 lb) for the female. Its length is up to 175 cm for the male, and 148 cm for the female.
2 This feature was noted by Audubon:
'In Dr Richardson's introduction to the second volume of the *Fauna Boreali-Americana*, we are informed that the Pelecanus Onocrotalus (which is the bird now named P. *Americanus*) [P. *erythrorhynchos* today] flies in dense flocks all the summer in the fur countries . . . My learned friend also speaks of the "long thin bony process seen on the upper mandible of the bill of this species," and although neither he nor Mr. [William] Swainson pointed out the actual differences otherwise existing between this

and the European species, he states that no such appearance has been described as occurring on the bills of the White Pelicans of the old Continent.' John James Audubon, *The Audubon Reader*, ed. Richard Rhodes (New York, 2006), p. 520.

3 In Srebarna, a nature reserve in Bulgaria, predation by wild boars is common. Wild boars destroy nests with eggs, or kill chicks.

4 During the 1970s and '80s Dalmatian pelicans at a number of key locations, including Mikri Prespa, Calmati Tuzlasi lagoon, Danube delta, Menderes delta, and Lake Skadar, were persecuted in this way. Bryan J. Nelson, *Bird Families of the World: Pelicans, Cormorants and their Relatives* (Oxford, 2005), p. 265.

5 Allopatric speciation, or geographic speciation, occurs when biological populations of the same species become changed, or isolated from each other to the extent that prevents genetic interchange.

6 Nelson, *Bird Families*, p. 250.

7 Ibid.

8 Ibid., p. 251. There are few differences between them in terms of trapping, catching and swallowing, but the pink-backed pelican may search for up to six times longer before trapping, and it feeds more frequently.

9 Ibid., p. 252.

10 V.E.M. Burke and L. H. Brown, 'Observations on the Breeding of the Pink-backed Pelican Pelecanus rufescens', *Ibis* (2008), pp. 499–512.

11 'Nagulu witnessed deaths of 5 nestlings and an adult during 6 hours in May 1980, when temperature rose to 42°C. Similarly in 1982, temperature in high 30s between 2 and 4 April led to 1, 5, and 15 deaths, all between 1300 and 1500 hours. Young birds dropped from trees and were unable to reach receded water.' Nelson, *Bird Families*, p. 258.

12 Ibid., p. 278.

13 'Mystery bird: Peruvian pelican, Pelecanus thagus', www.theguardian.com, 23 March 2012.

14 Alfredo Begazo, 'The Brown and Peruvian Pelicans', www.10000birds.com, 8 October 2015.

15 Ibid.

16 David Cameron Duffy, 'Competition for Nesting Space Among Peruvian Guano Birds', *Auk*, 100 (1983), p. 685, available at https://sora.unm.edu.

17 In Chile, photographic records show young Peruvian pelicans preying on juvenile Peruvian diving-petrels and on grey gulls off the coast of the iv Region de Coquimbo. See James A. Cursach et al., 'Presence of the Peruvian *Pelecanus thagus* in Seabird Colonies of Chilean Patagonia', *Marine Ornithology*, 44 (2015), p. 30, available at www.marineornithology.org, accessed 12 February 2017.

18 Some ornithologists say that the brown pelican is the only pelican that plunge dives; others disagree, saying that the Peruvian pelican also plunge dives. See, among others, 'Pelican', www.worldanimalfoundation.org, accessed 12 February 2017 and 'Mystery bird: Peruvian pelican, Pelecanus thagus', www.theguardian.com, 23 March 2012. What we can say with certainty is that the Peruvian pelican uses both methods.

19 Carlos B. Zavalaga et al., 'Patterns of gps Tracks Suggest Nocturnal Foraging by Incubating Peruvian Pelicans (Pelecanus thagus)', *plos one*, 6 (2011), http://journals.plos.org, accessed 12 February 2017.

Select Bibliography

Adcock, Fleur, and Jacqueline Simms, eds, *The Oxford Book
    of Creatures* (Oxford, 1997 edn)
Andrews, Ted, *Animal Speak: The Spiritual and Magical Powers
    of Creatures Great and Small* (St Paul, MN, 1993)
Armstrong, Edward A., *The Life and Lore of the Bird: In Nature,
    Art, Myth, and Literature* (New York, 1975)
Attenborough, David, *The Life of Birds* (London, 1998)
Audubon, John James, *The Audubon Reader*, ed. Richard Rhodes
    (New York, 2006)
Base, Graeme, *My Grandma Lived in Gooligulch* (Camberwell,
    Victoria, 1983)
Birkhead, Tim, *The Wisdom of Birds* (New York, 2008)
Bodkin, Frances, and Lorraine Robertson, *D'harawal Seasons
    and Climatic Cycles* (Sydney, 2008)
*The Book of Beasts: Being a Translation from a Latin Bestiary
    of the Twelfth Century*, ed. T. H. White (New York, 1984)
Borges, Jorge Luis, *The Book of Imaginary Beings*
    (London, 2002)
Brewer's Dictionary of Phrase and Fable, revd by Adrian Room
    (London, 2001)
Bright, Michael, *Beasts of the Field: The Revealing Natural History
    of Animals in the Bible* (London, 2006)
Brown, Augustus*, Why Pandas Do Handstands and Other Curious
    Truths about Animals* (New York, 2006)
Chitty, Susan, *That Singular Person Called Lear* (Stroud, 2007)

Colegate, Isabel, *A Pelican in the Wilderness: Hermits, Solitaries and Recluses* (New York, 2002)

The Complete Dictionary of Symbols in Myth, Art and Literature, ed. Jack Tresidder (London, 2004)

Cooper, J. C., *Symbolic and Mythological Animals* (London, 1992)

Couzens, Dominic, *Birds by Behaviour* (London, 2003)

Dahl, Roald, *The Giraffe and the Pelly and Me* (London, 2001 edn)

de Fournival, Richard, *Master Richard's Bestiary of Love and Response*, trans. Jeanette Beer (Berkeley and Los Angeles, CA, 1986)

Deniger, Lynda Wurster, *Patti Pelican and the Gulf Oil Spill* (Los Angeles, CA, 2011)

Fernando, Diana, *Alchemy: An Illustrated A to Z* (London, 1998)

ffrench, Richard, *A Guide to the Birds of Trinidad and Tobago* (Ithaca, NY, and London, 2012)

Forbes, Scott, *A Natural History of Families* (Princeton, NJ, and Oxford, 2005)

Fraser, Ian, and Jeannie Gray, *Australian Bird Names: A Complete Guide* (Collingwood, 2013)

Gibson, Graeme, *The Bedside Book of Birds: An Avian Miscellany* (London, 2005)

Gulpilil, David, *Stories of the Dreamtime*, compiled by Hugh Rule and Stuart Goodman (Sydney, 1979)

Haeffner, Mark, *Dictionary of Alchemy* (London, 1994)

Hahn, Daniel, *The Tower Menagerie* (London, 2003)

Harding, Mike, *A Little Book of Miracles and Marvels* (London, 2008)

Hart-Davis, Duff, *Audubon's Elephant* (London, 2004)

Harter, Walter, *Birds in Fact and Legend* (New York, 1979)

Hyman, Susan, *Edward Lear's Birds* (New York, 1980)

Ingpen, Robert, *The Voyage of the Poppykettle* (Adelaide, 1980)

—, *The Unchosen Land* (Adelaide, 1981)

—, with commentary by Sarah Mayor Cox, *Pictures Telling Stories: The Art of Robert Ingpen* (South Melbourne, 2004)

Joy, Charles R., *The Animal World of Albert Schweitzer* (Hopewell, NJ, 1950)

Lambourne, Maureen, *The Art of Bird Illustration* (London, 2005)

Lawrence, Michael, *The Poppykettle Papers*, illus. Robert Ingpen (London, 1999)

Lear, Edward, *The Complete Nonsense of Edward Lear*, ed. and intro. Holbrook Jackson (London, 1984)

—, *The Nonsense Verse of Edward Lear*, illus. John Vernon Lord (London, 1986)

Levi, Peter, *Edward Lear: A Biography* (London, 1995)

McNamee, Gregory, *Aelian's The Nature of Animals* (San Antonio, TX, 2011)

Mann, Charles C., *1491: New Revelations of the Americas Before Columbus* (New York, 2011)

May, John, and Michael Martin, *The Book of Beasts* (Feltham, 1982)

Mercatante, Anthony S., *Encyclopedia of World Mythology and Legend* (Frenchs Forest, NSW, 1988)

Metford, J.C.J., *Dictionary of Christian Lore and Legend* (London, 1983)

Mountford, Charles P., *The Dreamtime: Australian Aboriginal Myths* (Adelaide, 1965)

Nelson, J. Bryan, *Bird Families of the World: Pelicans, Cormorants and their Relatives* (Oxford, 2005)

Noakes, Vivien, *Edward Lear* (London, 1985)

Nozedar, Adele, *The Secret Language of Birds: A Treasury of Myths, Folklore and Inspirational True Stories* (London, 2006)

O'Grady, Gladys Y., and Terence Lindsey, *Australian Birds and their Young* (Stanmore, NSW, 1979)

Oliver, Narelle, *The Best Beak in Boonaroo Bay* (Port Melbourne, 1995 edn)

Page, Michael, and Robert Ingpen, *Encyclopedia of Things That Never Were* (London, 1997 edn)

Pickering, Fran, *The Illustrated Encyclopedia of Animals in Nature and Myth* (London, 2006)

Pinney, Roy, *The Animals in the Bible: The Identity and Natural History of All the Animals Mentioned in the Bible* (Philadelphia, PA, 1964)

Pocket Factfile of Birds: 200 Birds From Around the World, ed. Graham Bateman (Collingwood, Victoria, 1996)

Pope, Lawrence, *The Teeth of Beasts* (Brisbane, 2012)
Reed, A. W., *Aboriginal Myths: Tales of the Dreamtime*
    (Frenchs Forest, 1978)
—, *Aboriginal Legends: Animal Tales* (Sydney, 1978)
—, *More Aboriginal Stories of Australia* (Sydney, 1980)
Reilly, Pauline, *Pelican* (Kenthurst, NSW, 1997)
Robinson, Roland, *Aboriginal Myths and Legends* (Melbourne, 1966)
Schweitzer, Albert, *The Story of My Pelican* (London, 1964)
Scott, Bill, *Pelicans and Chihuahuas and Other Urban Legends*
    (St Lucia, 1996)
Soper, Tony, *Oceans of Birds* (Newton Abbot, Devon, 1989)
Sparks, John, *The Discovery of Animal Behaviour* (London, 1982)
Stap, Don, *Birdsong* (New York, 2005)
Stott, John, *The Birds: Our Teachers. Collector's Edition:*
    Essays in Orni-theology (Oxford, 2007)
Stutchbury, Bridget, *The Private Lives of Birds* (New York, 2010)
Talbot, Wayne, and Greg McKee, *The Great Wungle Bungle Aerial*
    Expedition (Richmond, Victoria, 1992)
Tate, Peter, *Flights of Fancy: Birds in Myth, Legend and Superstition*
    (London, 2007)
Thiele, Colin, *Storm Boy*, illus. Robert Ingpen (Sydney, 1992 edn)
Toperoff, Shlomo Pesach, *The Animal Kingdom in Jewish Thought*
    (Northvale, NJ, 1995)
Trounson, Donald, and Molly Trounson, *Australian Birds:*
    A Concise Photographic Field Guide (Seaford, Victoria, 2004)
Tudge, Colin, *The Secret Life of Birds* (London, 2009)

Acknowledgements

My thanks go to David, whose editing skills and help with referencing were invaluable; I love you 'to the moon and back'. So, too, to Rhys, Cara, Leaf, Harry, Pigeon and Yeti. To my Dumpling Sisters: Dianne, Leanne, Joanne and Kate; your laughter makes me soar! And for Kim, you are one of the bravest people I know; thank you for your love and friendship. To Michael Leaman and the staff at Reaktion, you are wonderful, as always. A huge thank you to the magnificent editorial team. This book could not have been written without the ability to draw on previous research undertaken by ornithologists and zoologists, writers and scholars, poets, artists and film-makers. To each and every one of you, thank you for shining your particular light on the pelican. A special 'thank you' to J. Bryan Nelson, whose monumental work *Bird Families of the World: Pelicans, Cormorants and their Relatives* was my 'go to' during the early stages of my research, and to Colin Thiele, whose story *Storm Boy* broke my heart many moons ago, and on each re-reading, never fails to reduce me to tears. Oh I did yearn for my own Mr Percival! I have tried to acknowledge all sources, but if I have overlooked anyone, my heartfelt apologies.

Photo Acknowledgements

The author and publishers wish to express their thanks to the following sources of illustrative material and/or permission to reproduce it. Some locations are also supplied here for reasons of brevity.

Academy of Fine Arts, Vienna: p. 75; photo akg-images / Mondadori Portfolio / Sergio Anelli: p. 66; photo akg-images / Mondadori Portfolio / Paolo e Federico Manusardi: p. 104; photo akg-images/Sputnik: p. 115; from John James Audubon, *The Birds of America* . . . (London, 1827–30): pp. 53, 137, 141; photo Paolo Auilar/EPA/Rex Features/Shutterstock: p. 150; photo Armando Babani/EPA/Rex Features/Shutterstock: p. 165; photos Barry Bland/Nature Picture Library: pp. 22, 37; photo Frederick William Bond (reproduced by kind permission of the Zoological Society of London): p. 91; photo Neil Bowman/FLPA/imageBROKER/Rex Features/Shutterstock: p. 27; from Hieronymus Brunschwig, *Liber de arte Distillandi de Compositis* (Strassburg, 1512): p. 84; photo Paul Buck/EPA/ Rex Features/Shutterstock: p. 158; Cooper Hewitt, Smithsonian Design Museum, New York (Open Access): p. 60; photo Richard Cummins/ robert harding/Rex Features/Shutterstock: p. 129; photo Tui de Roy/ Nature Picture Library: p. 47; Dallas Museum of Art, TX, reproduced by kind permission: pp. 122, 128; photo dieKleinert/Alamy Stock Photo: p. 139; photo digitalg/2018 iStock International, Inc: p. 33; from Daniel Elliott, 'A Monograph of the Genus Pelecanus', in *The Proceedings of the Zoological Society of London* (London, 1869): p. 12; from Leopold Joseph Fitzinger, *Bilder-Atlas zur wissenschaftlich-populären Naturgeschichte der Vögel in ihren sämmtlichen Hauptformen* (Vienna, 1864): p. 35; from Henry

O. Forbes, 'Notes on Molina's Pelican (*Pelecanus thagus*)', in *The Ibis, A Quarterly Journal of Ornithology*, II/3 (July 1914): p. 32; photo FLPA/ Bernd Rohrschneider/Rex Features/Shutterstock: p. 123; photo Getty Images/REPORTERS ASSOCIES: p. 168; photo Ann Marie Gorden (U.S. Coast Guard)/ U.S. Fish and Wildlife Service: p. 159; photo Paul Hobson/ Nature Picture Library: p. 26; photo imageBROKER/Rex Features/ Shutterstock: p. 164; J. Paul Getty Museum (Open Access): pp. 70, 83, 112; from Jon Jonston, *Historiae naturalis de avibus . . .* (Amsterdam, 1657): p. 87 (photo Mary Evans Picture Library); from [Charles Knight], *Charles Knight's Pictorial Museum of Animated Nature, and Companion for the Zoological Gardens*, vol. I (London 1856–68): pp. 16, 21; from Andrea Laurentio [André du Laurens], *Historia anatomica humani corporis et singularum eius partium multis controversiys et observationibus novis illustrata Andrea Laurentio . . .* (Frankfurt, 1600): p. 87; the Library of Congress, Washington, DC (Prints and Photographs Division): pp. 95, 98 (Historic American Buildings Survey), 119, 136, 174; photo Antoine Louchart, from A. Louchart, Nicolas Tourment and Julie Carrier, 'The Earliest Known Pelican Reveals 30 Million Years of Evolutionary Stasis in Beak Morphology', in *Journal of Ornithology*, CXLII/1 (January 2011): p. 13; photo Dian McAllister/Nature Picture Library: p. 25; from Saverio Manetti, *Ornithologia, methodice digesta atque iconibus aeneis ad vivum illuminatis ornate* (Florence, 1767–76): p. 42; photo Mary Evans Picture Library: p. 106; photo © Medici/Mary Evans Picture Library: p. 80; Metropolitan Museum of Art (Open Access): pp. 63, 65, 72, 76, 90, 105, 133; photograph Metropolitan Museum of Art (Open Access): p. 54; photo Morguefile: p. 176; National Gallery of Art, Washington, DC (Open Access): p. 109; photo National Gallery of Art, Washington, DC (Open Access): p. 108; National Gallery of Prague: p. 120; National Portrait Gallery, London: pp. 88, 89; from *The Natural History of Birds: From the works of Oliver Goldsmith, and all the best authors, antient & modern . . .* (London, 1821): p. 132; photo © Ondrooo /2018 iStock International Inc.: p. 172; private collections: pp. 23, 57 (photo Universal History Archive/Getty Images); Rijksmuseum, Amsterdam (Open Access): pp. 29, 44, 110, 111, 126; photographs Rijksmuseum, Amsterdam (Open Access): pp. 38, 67, 68, 143; from H. J. Ruprecht, *Wand-atlas für den*

Index

Page numbers in *italics* refer to illustrations